AESS Interdisciplinary Environmental Studies and Sciences Series

Series Editor

Wil Burns, Forum for Climate Engineering Assessment, School of International Service, American University, Washington, DC, USA

Environmental professionals and scholars need resources that can help them to resolve interdisciplinary issues intrinsic to environmental management, governance, and research. The AESS branded book series draws upon a range of disciplinary fields pertinent to addressing environmental issues, including the physical and biological sciences, social sciences, engineering, economics, sustainability planning, and public policy. The rising importance of the interdisciplinary approach is evident in the growth of interdisciplinary academic environmental programs, such Environmental Studies and Sciences (ES&S), and related 'sustainability studies.'

The growth of interdisciplinary environmental education and professions, however, has yet to be accompanied by the complementary development of a vigorous and relevant interdisciplinary environmental literature. This series addresses this by publishing books and monographs grounded in interdisciplinary approaches to issues. It supports teaching and experiential learning in ES&S and sustainability studies programs, as well as those engaged in professional environmental occupations in both public and private sectors.

The series is designed to foster development of publications with clear and creative integration of the physical and biological sciences with other disciplines in the quest to address serious environmental problems. We will seek to subject submitted manuscripts to rigorous peer review by academics and professionals who share our interdisciplinary perspectives. The series will also be managed by an Editorial board of national and internationally recognized environmental academics and practitioners from a broad array of environmentally relevant disciplines who also embrace an interdisciplinary orientation.

More information about this series at http://www.springer.com/series/13637

Valerie Banschbach · Jessica L. Rich
Editors

Pipeline Pedagogy: Teaching About Energy and Environmental Justice Contestations

Editors
Valerie Banschbach
Office of the Provost
Gustavus Adolphus College
Saint Peter, MN, USA

Jessica L. Rich
Independent Scholar
Boulder, CO, USA

ISSN 2509-9787 ISSN 2509-9795 (electronic)
AESS Interdisciplinary Environmental Studies and Sciences Series
ISBN 978-3-030-65978-3 ISBN 978-3-030-65979-0 (eBook)
https://doi.org/10.1007/978-3-030-65979-0

© Springer Nature Switzerland AG 2021
This work is subject to copyright. All rights are reserved by the Publisher, whether the whole or part of the material is concerned, specifically the rights of translation, reprinting, reuse of illustrations, recitation, broadcasting, reproduction on microfilms or in any other physical way, and transmission or information storage and retrieval, electronic adaptation, computer software, or by similar or dissimilar methodology now known or hereafter developed.
The use of general descriptive names, registered names, trademarks, service marks, etc. in this publication does not imply, even in the absence of a specific statement, that such names are exempt from the relevant protective laws and regulations and therefore free for general use.
The publisher, the authors and the editors are safe to assume that the advice and information in this book are believed to be true and accurate at the date of publication. Neither the publisher nor the authors or the editors give a warranty, expressed or implied, with respect to the material contained herein or for any errors or omissions that may have been made. The publisher remains neutral with regard to jurisdictional claims in published maps and institutional affiliations.

This Springer imprint is published by the registered company Springer Nature Switzerland AG
The registered company address is: Gewerbestrasse 11, 6330 Cham, Switzerland

For each of the communities disrupted by pipelines

Foreword

In most higher education I can think of in the U.S., the students learn little about where the energy they depend on comes from. Bizarrely, there's little coverage in the courses they take of where people get the power for their lights, heat, cooling, hot water, charging, and transportation. There's almost no curricular content about where the energy comes from for the manufacturing and materials needed for buildings, commercial products, and everyday items.

Higher education must educate about the sources of energy, energy infrastructure, and energy distribution. Every choice that college graduates make about their careers is only possible if the energy exists to undergird their intended pathways. How disconcerting that the educational institutions that provide markers of professional potential and competence in an industrial country like the U.S. do so little to educate about the most basic forces at play. Even within specific majors that train students to work in the energy sector, these majors are often bereft of substantive learning about moral, justice, social, and historical issues that are inextricably connected to the procurement of energy.

Higher education's failure in energy education is troubling in an industrial economy like the U.S. The major industrial sources of energy are coal, oil, and gas. They are dirty and non-renewable. They are obtained by extracting buried fossil fuels, refining them, and burning them to be consumed. The scale of pollution and desecration of extraction, refining, and burning is traumatic for many people who live and work nearby—and should be morally disturbing to everyone who benefits from the energy. The intensification of fossil fuel burning increases concentrations of greenhouse gases in the atmosphere, which destabilizes the climate system. Climate change risks, including increased frequencies of extreme weather events and expanding disease vectors, pose extraordinary threats.

Every decision a student in the U.S. might make mobilizes fossil fuel energy, whether they make these decisions as a reformer, consumer, citizen, family member, advocate, entrepreneur, politician, immigrant, friend, or scientist. As teachers on campuses, no matter what discipline, we have to make explicit where energy comes from. How much power does it take to make digital and online technologies work in the classroom? What sources of energy made British imperialism possible, which fueled the arts, literature, and philosophy that is so thoroughly covered in the curriculum? What industries subsidize engineering and geoscience fields for the sake of prioritizing certain types of energy? What communities sacrifice silently so that other persons can enjoy privileged access to affordable energy?

The aforementioned questions are about morality and justice. It is no accident that the choices of students and alumni mobilize fossil fuel energy. In the U.S., the capacity to extract, refine, and burn fossil fuels is rooted in oppression. The U.S. had to violently dispossess Indigenous peoples of their lands to make way for the 'space' needed to construct fossil fuel infrastructure, from mining basins to gas stations. Indigenous, African-American, Asian-American, and Latinx peoples, among other groups, have suffered for generations at the hands of fossil fuel industries. They have been subject to disproportionate levels of pollution, sexual violence, police and military brutality, and exploitative labor practices tied to these industries. The U.S. oppressed communities of people of color for the sake of building energy forces that could empower white men and wealthy white communities to feel secure in their freedoms.

Melanie Yazzie has lifted up the voices of energy justice advocates in the Navajo Nation who have resisted harmful uranium and coal extraction industries. Citing John Redhouse, who wrote in his biography about the border town of Farmington, New Mexico:

> The energy boom of the 50s and 60s brought the boomers and that's when Indian killing became a regular sport in Farmington. They would kill you just because you were Indian. So [we] grew up fighting during that particularly violent period. We had to fight back to survive … and while we were fighting for our lives, we realized the supreme irony that most of the energy that made Farmington a boomtown came from the nearby … Indian reservations. And that much of the water in the rivers which flowed through our tribal lands were used for regional energy development which benefited not only the area boomers but large off-reservation, non-Indian populations in big cities. … Oh my god, we were a colony, an exploited energy and water resource colony of the master race. (Yazzie [2], 33)

The exclusion of energy within higher education disappears the reality of immorality and injustice in the energy sector that Yazzie and Redhouse describe. In the absence of through coverage of the U.S. energy sector, what decisions, then, are teachers preparing their students to be able to make in the future? Who are they training their students to be?

Oil and gas pipelines are the conduits of fossil fuel energy across the U.S. supply chain. If one does an internet search for maps of pipelines, one will find they are among the most ubiquitous features of American industrial landscapes, whether they are underground, underwater, or hidden from public attention aboveground. But for communities who are most affected by pipelines, they are not 'under' anything or 'hidden' at all. Pipelines transport dangerous fluids through sacred and biodiverse places. Pipelines traverse the locations where communities live, work, and play.

In 2016, numerous Oceti Sakowin relatives fought to protect their lands and waters from being desecrated by the construction of the Dakota Access Pipeline, which would transport oil from the Bakken extraction region. There wasn't just one issue of concern, however. The pipeline threatened the water quality of the Missouri River, which members of the Standing Rock Sioux Tribe and Cheyenne River Sioux Tribe depend on. The pipeline construction desecrated sacred sites.

Yet the Dakota Access Pipeline was only possible as a project in the first place because of U.S. treaty violations that dispossessed Oceti Sakowin peoples of their homelands, making it challenging for them to exercise self-determination. For some time, Tribes near to the Bakken region have been deeply concerned about documented violence related to the oil boom, including sexual violence. Tribal Executive Board member of a nearby tribe (Ft. Peck), Marva Chapman, has stated that "The bottom line is a pipeline is contaminating to our water and to our people [1]."

The immorality and injustice of an educational system that ignores or masks these daily realities of violence is staggering. Another insidious feature is that the U.S. has suppressed Indigenous peoples' cultures—including Indigenous education systems. Speaking of my own background, Potawatomi and other Anishinaabe peoples have educational traditions that never tried to hide where our energy came from. Throughout peoples' lives, they discussed the importance of understanding the flows of wind, water, plants, animals, and soils. Understanding required humility. People were taught about the risks of failing to respect certain energy flows, and the importance being responsible in how one harnesses power. Anishinaabe education is transparent about where energy comes from, recognizing that people cannot pursue their lifeways morally and justly without sobering knowledge of sources of energy.

Pipeline Pedagogy is just the book needed for higher education to actually begin to have curriculum about how energy is made. Pipelines are among the visceral reminders of suffering and disempowerment for diverse communities impacted by U.S. industrialism. At the same time, for other communities—especially the economically privileged ones—pipelines are not even an afterthought as someone goes about their daily activities. No matter who students seek to become in their careers, they need to know about where energy comes from, in all that that means.

Lansing, MI, USA
June 2020

Kyle Whyte

References

1. Chapman, Marva (2019) Quote from Drew Novak, Fort Peck: The Fear Next Door: The Man Camp Connection. Native News J. Nativenews.jour.umt.edu/2019/fort-peck/ exact date in 2019 unknown.
2. Yazzie, Melanie K. (2018) "Decolonizing Development in Diné Bikeyah: Resource Extraction, Anti-Capitalism, and Relational Futures." Environ Soc 9(1): 25–39. Citing Redhouse, John. 2014. *Getting It Out of My System*. Self-published.

Preface and Acknowledgments

At the time of submitting this book for publication, news about pipeline controversies has been sweepingly displaced by news about Covid-19. We face an urgent need to focus on coping with life in a suddenly changed society. The disruption caused by the global pandemic was unthinkable just a year ago when we began the synthesis of the chapters in this volume. Now we anticipate never returning to the status quo of our economies and communities, even after a vaccine is deployed. What will this mean for environmental contestations? What is the future of energy production? With the price of oil sinking and fossil fuel usage declining, the environment has reaped some immediate benefits from human catastrophe. Will the future extend these environmental gains as we remake our ways of living? For now, the pipelines are still being built while the pipeline protests temporarily subside.

    This book is aimed at those who care to engage in learning and teaching about issues of great importance to communities: energy and environmental justice contestations. Faculty who are aiming to introduce students to the complex, multifaceted, and charged negotiations and confrontations among stakeholders in fossil fuel pipeline-impacted communities will find many innovative ideas presented by our authors. Faculty who wish to offer students community engagement must bring them into environments where civil discourse and fact-based argumentation may not prevail. Our chapter authors describe how they have leveraged those settings toward learning that crosses disciplinary boundaries and provides true practical value to students. Some of our chapter authors do not approach pipeline controversies as academic case studies. They share their work as activists and community organizers, and their work teaching students and community members to become activists, providing pragmatic strategies that suit the many fights facing environmentalists. We hope those fights let up. We hope that a post-Covid-19 world might see a rethinking of values that have led to the relentless push for fossil fuel infrastructuring. In the meantime, this volume can be a useful guide toward educating about environmental justice contestations and building dialogue in the midst of conflict.

Acknowledgments

This book arose from a session at the Association for Environmental Studies and Sciences (AESS) national conference in 2018 where Valerie Banschbach invited faculty, including Jessica Rich, to present work they were doing to educate about pipeline controversies. We would like to thank AESS for providing a vibrant and collegial home for interdisciplinary Environmental Studies and Sciences scholars. Without AESS, this project would not have been born. We are much obliged to all of the faculty around the world who are teaching about environmental justice contestations and engaging their students in learning about these difficult battles.

Valerie Banschbach was supported by her colleagues and community partners in teaching about the Mountain Valley Pipeline as Chair of the Environmental Studies Program at Roanoke College in Salem. Virginia. Without her colleagues and community partners, she would not have become engaged with the pipeline fight. She also appreciates the support of her colleagues in the Provost Office at Gustavus Adolphus College in Minnesota. They have been patient with her busy-ness as she worked to bring this volume to fruition while also working to transition in her first year of the Dean of Sciences and Education and Associate Provost position at Gustavus.

Jessica Rich began *Pipeline Pedagogy* while teaching in the Department of Communication and Media and the Environmental Studies Program at Merrimack College in North Andover, Massachusetts. She would like to offer her deep gratitude to her department colleagues at Merrimack–Samantha Bruno, Lisa Glebatis Perks, Andrew Tollison, Jacob Turner, and Melissa Zimdars–for providing a supportive and intellectually engaging space to pursue the interdisciplinary research necessary to complete this endeavor. In addition, she feels fortunate for the opportunity to teach the students in her Environmental Communication and Environmental Justice courses, whose compassion and commitment to a more sustainable future continue to motivate her work. Jessica also wishes to thank her parents, Mary Ann and Richard Rich; her sisters, Elizabeth Rich and Dina Rich; and her husband, Michael Willis, for their encouragement throughout this project.

We both would like to thank the editor of the AESS Branded Book Series, Dr. Wil Burns, and the staff at Springer, in particular Nel VanderWerf and Paul Roos, for their kind assistance and tolerance of our needs to adjust our timetable.

Saint Peter, MN, USA	Valerie Banschbach
Boulder, CO, USA	Jessica L. Rich

Contents

1 Introduction to Pipeline Pedagogy: Teaching About Energy
and Environmental Justice Contestations 1
Valerie Banschbach and Jessica L. Rich

Part I Teaching and Learning Across Disciplines

2 The Pipeline Case: Cross-Disciplinary Learning
and Pedagogical Lessons from the Mountain Valley Pipeline 13
Andreea Mihalache-O'Keef, Katherine O'Neill, Robert S. Emmett,
Marwood Larson-Harris, and Valerie Banschbach

3 Learning to Undermine a Pipeline: A Multi-logue
on Encounters with Vermont's Addison Natural Gas Project 33
Julie Macuga, Ingrid L. Nelson, Rachel Smolker, Trish O'Kane,
and Brian Tokar

4 We Are Teachers and Learners Together: Cross-Disciplinary
Lessons from the Pilgrim Pipelines Dispute 55
Lisa Jordan

Part II Tools and Methods for Teaching Pipeline Controversies

5 The Stop PennEast Pipeline Fieldwork Project: Teaching
Students to Apply Fieldwork Methods to Studying a Natural
Gas Pipeline Opposition Movement 75
Michael J. Brogan

6 Extractive Messaging: A Critical Communicative Approach
to Pipeline Pedagogy 91
Jessica L. Rich

Part III Mobilizing Pipeline Politics

7   **Mountain Valley Pipeline: A Case Study in Local Resistance
    and Mobilization** 107
    Diana Christopulos

8   **Linking Sovereignty, Local Environments, and Climate Justice
    Through Pipeline Pedagogy** 141
    Theodor Gordon, Corrie Grosse, and Brigid Mark

Index 157

Chapter 1
Introduction to Pipeline Pedagogy: Teaching About Energy and Environmental Justice Contestations

Valerie Banschbach and Jessica L. Rich

Abstract This chapter describes our motivations for assembling this volume, as well as the structure of the sections and key content of each chapter of the work. In this introduction to the volume we highlight the varied author perspectives, disciplinary and interdisciplinary, academic and activist, that illuminate the usefulness of pipeline controversies, and other kinds of energy contestations, in teaching that connects our students with the communities surrounding their colleges. In this chapter, we introduce the means by which this volume considers issues of environmental justice, legitimacy, and inclusion in terms of types of knowledge which count as evidence and in terms of power relations in the struggle over fossil fuel pipelines.

Keywords Fossil fuel pipelines · Community-engaged pedagogy · Environmental justice · Transdisciplinary · Energy development · Infrastructuring

1.1 Why Teach About Fossil Fuel Pipelines?

The proliferation of pipelines to transport oil and natural gas represents a major area of contestation in the landscape of energy development and, therefore, poses a need for education in formal and non-formal settings. Battles over energy pipelines pit private landowners, local community representatives, and environmentalists against energy corporations and industry supporters, sometimes drawing opposition and attention from well beyond the impacted regions [e.g., 7, 15, 10]. Stakeholders must navigate complex government regulatory processes, interpret technical and scientific reports, and endure lengthy and expensive court battles. As with other

---

V. Banschbach (✉)
Office of the Provost and Environmental Studies Program, Gustavus Adolphus College, Saint Peter, MN, USA
e-mail: vbanschbach@gustavus.edu

J. L. Rich
Independent Scholar, Boulder, CO, USA
e-mail: richjessica1@gmail.com

© Springer Nature Switzerland AG 2021
V. Banschbach and J. L. Rich (eds.), *Pipeline Pedagogy: Teaching About Energy and Environmental Justice Contestations*, AESS Interdisciplinary Environmental Studies and Sciences Series,
https://doi.org/10.1007/978-3-030-65979-0_1

forms of environmental injustice, the contentious construction of pipelines often disproportionately impacts communities of lower economic development, people of color, and Indigenous peoples Bullard [3]; Pellow [21]. Pipelines also pose potential short- and long-term health and safety threats [2, 25]. Natural gas fields such as the Marcellus Shale in the Appalachian Basin of the U.S. are experiencing a "boom," creating multifaceted challenges, necessitating transdisciplinary analysis [13]. With the expansion of energy pipelines carrying fracked oil and gas across the United States and abroad, the moment is ripe for teaching about pipeline projects and engaging students and community members in learning about methods for mobilization [11, 17, 19, 29]. Our volume examines opportunities, challenges, and interventions that campus–community and other community engagement produce in relation to pipeline development.

Teaching about energy contestations provides opportunities for a range of pedagogical approaches in Environmental Studies and Sciences, including the use of case studies, place-based teaching methods, community-based learning, and critical pedagogy. Case studies have proven efficacy in engaging students in critical thinking about multiple dimensions of controversies [4, 18, 31]. Energy contestations such as pipeline development also may strongly foreground the issue of "place" in ESS teaching, as some of the most feared impacts of pipeline projects stem from the fact that a single infrastructure project crosses dozens or even hundreds of different localities, potentially threatening the special features of each. Finley-Brook et al. [8] used the verb "infrastructuring" to "highlight the process of contested gas expansion" (p. 177) and analyze the issues within a critical energy justice framework. This follows from the use of "infrastructuring" more broadly, beyond natural gas infrastructuring, as a verb connoting many kinds of environmental change and concomitant contestations [1 p. 2]. Study of infrastructuring presents an opportunity to engage place-based learning methods [e.g., 26, 27]. Depending on the pipeline, or other specific forms of energy development, the controversy may afford an opportunity to teach about long-standing regional social justice issues that are triggered, such as the Mountain Valley Pipeline and the Atlantic Coast Pipeline raising issues of resource curse and sacrifice zones [20, 14]. Or the controversy may provide a chance to connect students with the broader landscape of thinking about energy development [e.g., 24]. Critical pedagogy, or "pedagogy of the oppressed" [9], can be applied to teaching about environmental justice by giving voice to Indigenous knowledge and elevating the perspectives of the oppressed in these power struggles, or viewing the conflicts through lenses of decolonization theory [6, 12, 16, 22, 28, 30].

1.2 Educational Contexts for Analyzing Energy Contestations

The chapters included in *Pipeline Pedagogy* showcase what the purposeful practice of campus–community engagement looks like in relation to environmental conflicts. As Putnam [23] noted two decades ago, the number of venues in which communities deliberate the day's critical issues continues to shrink in the United States.

The college classroom is one of the few remaining spaces in which individuals can openly debate energy justice and other divisive topics, a legacy of higher education's enduring commitment to academic freedom. Instructors, in our experiences, embrace the responsibility that their authority in the classroom entails and encourages students to engage in well-formed arguments amid a diversity of viewpoints. Campuses also are critical landscapes where students, faculty, and residents can engage with one another, with community speakers, and with administrators on our complex societal involvement and investments in fossil fuel industries. Pipeline contestations, as explored in this volume, provide multiple contexts for examining, analyzing, and debating energy development and demanding action for the environmental injustices that oil and gas bring into communities.

While community engagement can be fulfilling and rewarding for multiple stakeholders, it also is important for individuals entering into such projects to recognize the complex social challenges of campus–community partnerships, which can risk reproducing the complex power structures that divide university and residential populations [5]. The authors in the following chapters provide ample evidence that pipeline pedagogy increases faculty and student involvement in their off-campus communities. Pipeline pedagogy also creates opportunities for communities to benefit from the resources that universities and colleges provide. As in the classroom, however, off-campus spaces require similar, if not more vigilant, care on behalf of the campus community to engage and navigate the politics of pipeline controversies. Engaging in energy contestations inside and outside of the classroom necessitates that participants take care to emphasize the politics, rather than the partisanship, of pipeline controversies. In addition, instructors and students may call on community members, who volunteer their time as informal educators, guides, researchers, and public speakers, to provide uncompensated labor often performed on top of full-time jobs, child and elder care, and many times, at the sacrifice of self-care. It is critical that faculty and students who take on pipeline pedagogy projects recognize the relentless, and often overlooked, work that can accompany environmental protest.

Each project in this volume resulted from a commitment to learning and to relationship-building among multiple actors, including organizations, instructors, students, and members of pipeline opposition campaigns. It is our hope that the discussions included here provide inspiration, models, and guidance for readers across disciplines and organizational settings who are considering similar endeavors in their own communities. In addition, instructors taking on the issue of environmental justice often require students to reflect on their own political positionality in energy debates. The editors demand similar considerations of ourselves. Valerie and Jessica developed this project to call attention to a diversity of pedagogical projects and tools to teach about pipeline protests. While a strength of the volume is its interdisciplinary focus on environmental justice, a limitation is the scarcity of racial and ethnic diversity among the authors. This absence reflects the dominance of white perspectives across the fields of environmental studies and sciences (the primary source of author recruitment), as well as the campus and communities where much of the research was carried out. Communities of color continue to be targeted by energy companies and policymakers to bear the stress of oil and gas development

[3, 21]. As pipeline pedagogy research and community engagement move forward, more work is needed on the part of white editors, publishers, and authors working on these issues to seek out researchers, teachers, and activists who are people of color involved in pipeline campaigns.

1.3 Ways of Teaching About Energy Contestations

In outlining the status quo of energy education, as well as the political contexts in which teaching about energy contestations occurs, our goal is to situate this edited collection broadly within Environmental Studies and Sciences, and more specifically within energy education, while offering new insights gleaned from embracing interdisciplinary approaches. All of the chapter authors here present their own perspectives, using their distinctive skills and knowledge, to create experiments in pedagogy or community education. We purposely chose chapters whose authors come from different academic backgrounds (Earth Science, Ecology, Political Science, Communication, Sociology…) and whose authors come from different positions (faculty, students, community organizers, activists…) in order to provide a broad mix of lenses through which education about pipeline controversies may be viewed. The co-editors likewise come from different disciplinary backgrounds (V. Banschbach, Ecology; J. Rich, Communication) and different regions of the country but are united in being teacher–scholars who want to highlight the usefulness of pipeline controversies, and other kinds of energy contestations, for teaching that connects our students with the communities surrounding their colleges. They had not met until at a session Valerie organized for the Association for Environmental Studies and Sciences annual meeting (2018), which is a professional association dedicated to encouraging, "interdisciplinary understanding of environmental science, policy, management, ethics, history, and all of the other vital contributions of traditional disciplines in order to better understand the natural world and humans' relations with it." In that session, Valerie, Jessica, and others were impressed by both the range of approaches to teaching about pipelines and the range of pipeline projects and geographic regions represented (Mid-Atlantic, Northeast, Midwest, and West). It was clear from the enthusiasm of the audience and presenters that this work needed a chance to reach a broader audience. V. Banschbach and J. Rich worked together to organize a second Pipeline Pedagogy presentation session at the AESS (2019) annual meeting, allowing some of the original presenters to follow-up and allowing new presenters to join the mix.

In this volume, we wish to strongly consider legitimacy and inclusion in terms of types of knowledge which count as evidence and in terms of power relations in the struggle over these pipelines. Efforts toward co-production of knowledge, by people working together across sector boundaries, are needed to make progress on environmental challenges [32]. Therefore we include chapters authored not only by other teacher–scholars, but also by extra-academic actors, including practitioners, community organizers and activists. The authors' contributions respond to a number of critical questions that arise from the pedagogy of pipelines:

- How do pipeline curricula (formal and non-formal) create spaces that value local knowledge?
- What can public forms of expertise teach campus communities about the benefits and challenges of energy transitions?
- What lessons are learned about building meaningful campus–community partnerships in the midst of environmental conflict?
- How can Environmental Studies and Sciences support communities facing environmental injustices brought on by pipeline development?
- How can diverse forms of expertise, such as experiential and embodied knowledge, citizen science, and popular epidemiology, educate campus partners?

We organize the chapters based on three themes that emerge from the authors' work: Teaching and Learning Across Disciplines; Tools and Methods for Teaching Pipeline Controversies; and Mobilizing Pipeline Politics. First, "Teaching and Learning Across Disciplines" investigates efforts to educate and engage students, while connecting a range of disciplines, and educating the public on contentious issues of energy development. Next, "Tools and Methods for Teaching Pipeline Controversies" explores specific ways in which we can use the classroom as space for proactively engaging students in the study and work of pipeline debates. Finally, the chapters in "Mobilizing Pipeline Politics" focus on purposeful campus–community collaborations that have emerged from local responses to pipeline development, as well as outcomes of these actions.

1.3.1 Teaching and Learning Across Disciplines

Environmental Studies and Science are areas of study that have intentionally sought to distinguish themselves from the disciplines from which they originally sprung. Environmental Science connects different natural and physical sciences toward applying the scientific method toward understanding environmental problems. In a traditional framing, environmental scientists would seek understanding of sources of energy while applying that knowledge toward problems of how to gain efficiency during extraction and utilization of energy resources. Environmental Studies includes the sciences, but on par with social sciences and humanities, and seeks to join environmental understandings from across the spectrum of scholarly work. Environmental concerns do not lie within the boundaries of particular disciplines. In this section of the volume, authors work strongly toward crossing boundaries, both of academia and between academia and the rest of society.

In the "Teaching and Learning Across Disciplines" section of the volume, Chapter 2, Mihalache-O'Keef, O'Neill, Emmett, Larson-Harris and Banschbach (Environmental Studies, Roanoke College) explore the merits of cross-disciplinary learning spanning five ESS areas—Environmental Science, Earth Sciences, Environmental Humanities, Environmental Policy, and Environmental Ethics, as they present their teaching about the Mountain Valley Pipeline, a liquefied natural gas pipeline under construction in the Appalachian Mountains of SW Virginia. The Roanoke College chapter provides insight into the value of this pedagogy for the development of an environmental ethos in students as the faculty utilized formal assessment methods to understand the impacts of their program-wide teaching and program-level events about the Mountain Valley Pipeline on student environmental empathy and engagement with place. Like Mihalache-O'Keef et al., the authors of Chapter 3, Macuga, Nelson, Smolker, O'Kane, and Tokar, bring together different perspectives as authors, this time including a field organizer with 350Vermont and a faculty member who is a feminist political ecologist, interacting with the viewpoints of activists and educators. These authors take a multilogue approach, presenting a facilitated dialogue, transcribed and edited for print, focusing on their own "points of convergence and friction" in pedagogy as well as the broader question of how some of the greenest campuses in the country (e.g., some colleges and universities in Vermont) are interacting with the Addison Natural Gas Project, a fracked natural gas pipeline running from Alberta, Canada, into Vermont. To end this section, Jordan, Chapter 4, brings multiple disciplinary perspectives (Geographic Information Sciences, Public Health, and Environmental Studies) together in Community-Based learning coursework focused on the Pilgrim Pipelines that is heavily informed by her own experiences as a community member and activist. Jordan demonstrates that one person who is a teacher-scholar-activist can foster cross-disciplinary learning by working with students on projects that intersect with public health, Geographic Information Systems (GIS), and environment and society.

1.3.2 *Tools and Methods for Teaching Pipeline Controversies*

Engaging students in understanding controversial, local events requires careful consideration of how to deepen students' thinking about the events beyond the sound bites, headlines, and slogans on protest signs. In this section of the volume, our authors present useful techniques and approaches to pipeline pedagogy stemming from academic disciplinary methods. In Chapter 5, Brogan describes his use of fieldwork to enrich Political Science teaching focused on the PennEast Pipeline in Northeastern Pennsylvania and Northwestern New Jersey. Students use field surveys, canvassing, participant observations of meetings, survey methods, and content analysis to analyze multiple facets of the controversy and contextualize it within the theory of social movements. Rich, in Chapter 6, draws from the field of Communication to present discussion-based and multimedia activities as pedagogical methods that encourage students to analyze debates surrounding extractive activities. A public hearing over drilling and pipeline development in Colorado serves as a rich case study for students to analyze messages produced by local government, oil and gas companies, and the communities that resist pipeline projects and to examine how technical and local knowledge are valued differently in environmental decision-making processes.

1.3.3 Mobilizing Pipeline Politics

Pipeline controversies are inherently political in the sense that politics relates to achieving and exercising organized control over human communities. Seizing private property via eminent domain, disrupting that land and the surrounding landscape with heavy excavation and large-scale construction, transporting hazardous and/or potentially polluting fossil fuels, and monitoring the infrastructure periodically for maintenance equates to controlling the human communities impacted in each locale that a fossil fuel pipeline passes through. While these energy justice contestations are certainly political, they are not clearly partisan. In fact, strange bedfellows often come together in opposition to local projects. Conservatives fearful of government interference with private property rights may react as strongly as liberals who fear environmental damage by pipeline projects. In this third section of the volume, authors discuss the politics of particular pipeline projects and how those politics may be mobilized toward effective resistance by community education.

In Chapter 7, Christopulos, a community organizer, discusses her approach to mobilizing resistance toward the Mountain Valley Pipeline, in SW Virginia. Her coalition-building efforts focused on recruiting technical experts (scientists, economists, psychologists...) to speak at educational forums held in collaboration with faculty at local colleges, as well as grassroots mobilizing of volunteers from local small conservation organizations to speak and protest at public hearings. In Chapter 8, this section ends with the work of Gordon, Grosse, and Mark who work to center sovereignty of Indigenous perspectives in their pipeline pedagogy for resistance to the Line 3 tar sands pipeline in Minnesota. While the politics they mobilize focus on Native Nation sovereignty, these authors also provide strong cross-disciplinary grounding with insights from Environmental Studies, Sociology, Gender Studies, Anthropology, and Biology. They conclude with suggestions for activities that advance a pedagogy with the power to protect people, prevent pipelines, and promote justice.

References

1. Blok A, Nakazora M, Winthereik BR (2016) Infrastructuring environments. Sci Cult 25(1):1–22
2. Brown A (2018, January 9) Five spills, six months in operation: dakota access track record highlights unavoidable reality—pipelines leak. The Intercept. Retrieved from https://theintercept.com/2018/01/09/dakota-access-pipeline-leak-energy-transfer-partners/

3. Bullard RD (1994) Dumping in Dixie: race, class, and environmental quality. Westview Press, Boulder CO
4. Burns W (2017) The case for case studies in confronting environmental issues. Case Stud Environ 1(1):1–4. https://doi.org/10.1525/cse.2017.sc.burns01
5. Dempsey S (2010) Critiquing community engagement. Manag Commun Q 24(3):359–390
6. Estes N (2019) Our history is the future: standing rock versus the Dakota access pipeline, and the long tradition of Indigenous resistance. Verso Books, New York
7. Estes N, Dhillon J (eds) (2019) Standing with standing rock: voices from the #NoDAPL movement. University of Minnesota Press, Minneapolis
8. Finley-Brook M, Williams TL, Caron-Sheppard JA, Jaromin MK (2018) Critical energy justice in US natural gas infrastructuring. Energy Res Soc Sci 41:176–190
9. Freire P (1970) Pedagogy of the oppressed. Seabury Press, New York
10. Halper E (2017, July 14) A pipeline cutting through the iconic appalachian trail sparks a fight over natural gas expansion. *Los Angeles Times*. Retrieved from https://www.latimes.com/politics/la-na-pol-pipeline-appalachian-trail-201707-htmlstory.html
11. Hodges HE, Stocking G (2016) A pipeline of tweets: environmental movements' use of Twitter in response to the keystone XL pipeline. Environ Polit 25(2):223–247
12. Kahn R (2010) Critical pedagogy, ecoliteracy and planetary crisis. Peter Lang Publishing, New York
13. Kinne BE, Finewood MH, Yoxtheimer D (2014) Making critical connections through interdisciplinary analysis: Exploring the impacts of Marcellus shale development. J Environ Stud Sci 4:1–6
14. Lerner S (2010) Sacrifice zones: the front lines of toxic chemical exposure in the United States. MIT Press, Cambridge MA
15. Levin S, Woolf N (2016, Oct. 31) A million people 'check in' at standing rock on Facebook to support Dakota pipeline protesters. *The Guardian*. Retrieved from https://www.theguardian.com/us-news/2016/oct/31/north-dakota-acc esspipeline-protest-mass-facebook-check-in
16. Maldonado JK (2013) The impact of climate change on tribal communities in the US: displacement, relocation, and human rights. Clim Change 120(3):601–614
17. McAdam D, Schaffer Boudet H, Davis J, Orr RJ, Scott WR, Levitt RE (2010) 'Site fights': explaining opposition to pipeline projects in the developing world. Sociol Forum 25(3):401–427
18. National Center for Case Study Teaching in Science (2014). Retrieved from http://sciencecases.lib.buffalo.edu/cs/
19. Ordner JP (2017) Community action and climate change. Nat Clim Chang 7:161–163
20. Partridge MD, Betz MR, Lobao L (2013) Natural resource curse and poverty in Appalachian America. Am J Agric Econ 95(2):449–456
21. Pellow DN (2018) What is critical environmental justice? Polity, Cambridge
22. Porfilio BJ, Ford DR (eds) (2015) Leaders in critical pedagogy: narratives for understanding and solidarity. Sense Publishers, Rotterdam
23. Putnam RD (2001) Bowling alone. Simon and Schuster, New York
24. Sachs JD (2017) Building the new American economy: smart, fair and sustainable. Columbia University Press, New York
25. Siler-Evans K, Hanson A, Sunday C, Leonard N, Tumminello M (2014) Analysis of pipeline accidents in the United States from 1968 to 2009. Int J CritAl Infrastruct Prot 7(4):257–269
26. Sobel D (2004) Place-based education: connecting communities and classrooms. The Orion Society, Roxbury
27. Surface JL (2016) Place-based learning: instilling a sense of wonder. Publications of the rural futures institute 10 http://digitalcommons.unl.edu/rfipubs/10
28. Tuck E, Yang, KW (2012) Decolonization is not a metaphor. Decolonization: Indig, Educ Soc 1(1):1–40
29. Veltmeyer H, Bowles P (2014) Extractivist resistance: the case of the enbridge oil pipeline project in Northern British Columbia. Extr Ind Soc 1(1):59–68
30. Whyte K (2017) The Dakota access pipeline, environmental injustice, and U.S. colonialism. Red Ink: Int J Indig Lit, Arts, HumIties 19(1) https://ssrn.com/abstract=2925513

31. Yadav A, Lundeberg M, DeSchryver M, Dirkin K, Schiller NA, Maier K, Herreid CF (2007) Teaching science with case studies: a national survey of faculty perceptions of the benefits and challenges of using cases. J Coll Sci Teach 37(1):34–38
32. van der Hel, S (2016) New science for global sustainability? The institutionalisation of knowledge co-production in Future Earth. Environ Sci & Policy 61:165–175

Part I
Teaching and Learning Across Disciplines

Chapter 2
The Pipeline Case: Cross-Disciplinary Learning and Pedagogical Lessons from the Mountain Valley Pipeline

Andreea Mihalache-O'Keef, Katherine O'Neill, Robert S. Emmett, Marwood Larson-Harris, and Valerie Banschbach

Abstract We present a nested case study that highlights the pedagogical opportunities and challenges of implementing a single teaching case, the Mountain Valley Pipeline, across multiple courses and in co-curricular activities within an interdisciplinary Environmental Studies program. Our pedagogy focused on connecting environmental changes, including climate change, with political choices, institutions, ideologies, and cultural practices, as well as addressing issues of avoidance and disengagement among students. We first describe the various activities through which the pipeline case was incorporated in our courses, then offer a panoramic, program-level analysis of the advantages and challenges of this interdisciplinary pedagogical project, highlighting practical, theoretical, and community-building aspects.

Keywords Mountain Valley Pipeline · Community-engaged pedagogy · Appalachia · Environmental studies program · Nested case study · Energy transition

---

A. Mihalache-O'Keef (✉)
Department of Public Affairs and Environmental Studies Program,
Roanoke College, Salem, VA, USA
e-mail: mihalache@roanoke.edu

R. S. Emmett
College of Engineering, Virginia Tech, Blacksburg, VA, USA
e-mail: rsemmett18@vt.edu

K. O'Neill · M. Larson-Harris
Environmental Studies Program, Roanoke College, Salem, VA, USA
e-mail: oneill@roanoke.edu

M. Larson-Harris
e-mail: mdharris@roanoke.edu

V. Banschbach
Gustavus Adolphus College, Saint Peter, MN, USA
e-mail: vbanschbach@gustavus.edu

© Springer Nature Switzerland AG 2021
V. Banschbach and J. L. Rich (eds.), *Pipeline Pedagogy: Teaching About Energy and Environmental Justice Contestations*, AESS Interdisciplinary Environmental Studies and Sciences Series,
https://doi.org/10.1007/978-3-030-65979-0_2

2.1 Introduction

Environmental Studies positions itself as an "integrative interdiscipline" that transcends traditional academic boundaries and embraces complexity in ways that are both inclusive and coherent [12]. Yet, many educators on the ground remain faced with the fundamental challenge of integrating "credible and actionable interdisciplinarity" [12] within and across courses that may be housed within different academic departments and that embrace a range of disciplinary frameworks. In a survey conducted by the National Council for Society and the Environment, leaders of environmental studies programs in the United States identified the critical need for pedagogy that integrates human and natural systems (socio-environmental perspective) and emphasizes interdisciplinary problem-solving [17, 18]. Others, including Cronon [5], Clark et al. [4], and Lekan [8], have issued calls for a greater emphasis on incorporating interdisciplinary problem-solving approaches and integrative theory across ES curriculum. Deliberate program building—rather than conceptualizing interdisciplinarity—is necessary because "simply rubbing a few disciplinary sticks together may not produce interdisciplinary fire" [12].

Case studies offer one pedagogical approach for holistic exploration of complex issues in ways that "grapple with the interdependence of societal, ethical and environmental" concerns [3]. By focusing on a specific example that is grounded in both place and time, case studies offer an active, place-based approach that integrates theory and perspectives across disciplines, provides opportunities to practice interdisciplinary problem-solving skills, and even contributes to a heightened sense of agency in students [19]. Here, we present a nested case study that highlights the pedagogical opportunities and challenges of implementing a single teaching case, the Mountain Valley Pipeline, across multiple courses and in co-curricular activities within an interdisciplinary Environmental Studies program.

Our pedagogy focused on connecting environmental changes, including climate change, with political choices, institutions, ideologies, and cultural practices. Above all, we sought to address issues of avoidance and disengagement among students during the proposal and initial construction phases of the Mountain Valley Pipeline. In order to assess changes in student attitudes and environmental concerns, we conducted repeat surveys in each of the courses that featured the case study. This chapter uses this assessment data in conjunction with analysis and discussion of our teaching techniques to reflect on (1) the benefits and pitfalls of using multicourse but discrete "modules" as a pedagogy of engagement around contested energy projects; (2) benefits and challenges of employing an ongoing/unresolved issue as a means of combatting disengagement among undergraduate students; and (3) possibilities for developing a valid and reliable assessment metric for complex learning objectives such as student empathy and meaningful engagement with place.

2.2 The Mountain Valley Pipeline Case

At the time of our department-wide curricular efforts to teach about the Mountain Valley Pipeline (Fall of 2017), the proposed 300-mile natural gas distribution pipeline project was still in the final permitting phase. As outlined in the proposal, the pipeline would pump 2.5 billion cubic feet of liquid natural gas, daily, from fracked wells in the Marcellus Shale deposits located in West Virginia to a pumping junction in central Virginia. The 42-inch diameter pipe, unusually large for an interstate natural gas project, would extend across very steep mountain passes of up to 25% grade, the Jefferson National Forest, residential areas, and 51 miles of unstable karst topography, noted for its geological instability and abundant underground springs. The pipeline would make numerous headwater stream crossings, 986 total waterbody crossings, including three major aquifers and one Source Water Protection Area Twenty-two 22 federally listed threatened, endangered, or special concern species, were in the vicinity of the proposed pathway.

Many rural communities along the pipeline rely on well water, which was at risk of contamination during the construction phase as well as from leaks during operation. The highly fractured karst found in the mountain valleys presented a complex subsurface hydrology with multiple pathways for water flow that can be notoriously difficult to model and predict using conventional engineering approaches. Removal of vegetation and excavation on steep slopes increased the potential for erosion and for sediment contamination of first and second-order streams and for damage to the habitat for endangered aquatic species, including the Roanoke logperch. Unstable slopes, loosened by blasting and vegetation removal, also called into question the structural integrity of the pipeline itself along with the increased potential for leaks or even explosions resulting from landslides. Communities surrounded by dense forest with few access roads and no public water expressed concerns about the ability of first responders to combat wildfire and prevent the loss of life in the case of an emergency. Additionally, the aesthetic and tourism value of the land along the route includes iconic viewsheds of the Blue Ridge Parkway and the Appalachian Trail. The Mountain Valley Pipeline would result in homeowners losing part or all of their property under eminent domain seizures, including loss of some acreage on family farms, and the removal of hundreds of acres of forests.

Beyond the local community and environment, the pipeline engaged actors and raised issues about processes at other levels. It was an apt illustration for the FERC permitting process, the interests of the various stakeholders on the board, and the agency's responsiveness to citizen testimony. The federal government further participated in the issue through the invocation of eminent domain in carving out the path of the pipeline and through the environmental impact assessment process. The trajectory of the pipeline highlighted issues of power imbalances and environmental justice, as it was re-routed around wealthier communities. Questions were also raised about the type and location of the projected new jobs, who would be employed, and the hidden costs that would accompany the pipeline's construction and operation. The pipeline also carried the potential for public health problems: possible explosions, water contamination, and psychological stress to residents located in the vicinity of construction sites and the pipeline route itself. The Mountain Valley Pipeline project, hereafter referred to as MVP, was multidimensional and complicated, ensuring that any pedagogical response to it would necessarily be interdisciplinary. Our classes across the program would have to account for the sociopolitical, economic, cultural, and ecological context of the pipeline. As faculty and citizens, we had numerous questions to explore in order to understand this complex issue.

2.3 The Mountain Valley Pipeline in the Roanoke College Curriculum

Roanoke College is a private, liberal arts and sciences, primarily undergraduate institution that serves approximately 2,000 traditional undergraduate students. The mission of the College derives from an Evangelical Lutheran heritage but is strongly secular. Students at the College come from the East Coast of the United States with approximately half coming from the North Eastern United States and the remaining half from the Mid-Atlantic and South East. At the College as a whole, during the time period of our study, 256 RC students (12%) came from pipeline counties. The Mountain Valley Pipeline route runs within fourteen miles of the College campus.

By the fall of 2016, the MVP became visible in the Roanoke Valley and surrounding communities and several groups organized to give voice to those most likely to be affected, to disseminate research findings, and to develop resistance strategies. As community awareness continued to grow, Environmental Studies faculty incorporated discussion of the MVP in their courses. For instance, in the Environmental Public Policy class, students engaged in field observation of pipeline protests and used the MVP case to illustrate the policymaking process, stakeholder interests, and actor power. In upper-level Environmental Science courses, students examined the draft Environmental Impact Statement as examples of soil and water quality assessments and examined the role of cartography in public discourse using published maps. Students in Writing After the End of Nature engaged with a visiting member of the arts-activist Beehive Collective and produced a poster for the MVP public forum in Fall 2016.

The positive student response to these activities, including high levels of engagement and some student initiatives that went beyond the course requirements, prompted the ENVS faculty to become more intentional about the use of the MVP as a case study across the program curriculum. Our work spanned the inquiry-based general education curriculum at Roanoke College—the "Inquire" curriculum—and the natural sciences, social sciences, and humanities courses for the Environmental Studies Program (Fig. 2.1).

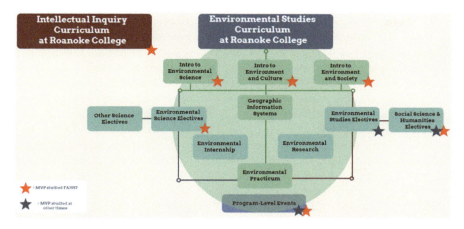

Fig. 2.1 Mapping the Mountain Valley Pipeline in the Environmental Studies Curriculum

2.3.1 Program-Level Goals and Theoretical Frameworks

In Fall 2017, the MVP case was incorporated into all three of the introductory courses in the Environmental Studies program (ENVS): Introduction to Environmental Science; Introduction to Environment and Culture; and Introduction to Environment and Society. The case was also used in upper-level Environmental Science, Social Sciences, and Humanities electives in the program and Inquire offerings. Several program-level events (e.g., public lectures) also had the MVP as their focus and we required students in Environmental Studies courses to attend a certain number of those events. The pedagogy used in the framing of the MVP case in different courses was aligned with learning goals for the Roanoke College Environmental Studies Program that were incorporated into each course as described here.

Environmental Studies Program Learning Goals (2016–2019):

1. Actively investigate roles and identities as citizens, consumers, and environmental actors
2. Use an interdisciplinary approach that integrates perspectives and/or methods
3. Demonstrate critical thinking, using quantitative and/or qualitative forms of analysis
4. Link theory and practice in order to understand the connections and how these are reflected "on the ground"
5. Work collaboratively in the design and implementation of research, applied projects, or creative works
6. Collect data using professional methods
7. Communicate effectively in writing, oral presentations, visual or digital media
8. Consider the significance of diversity in culture and circumstance to environmental issues

The MVP is a complex case that engaged a wide array of stakeholders, institutions, and values. The multidimensionality of the case—encompassing ethics, economics, politics, business, geology, and ecosystem impacts—allowed for multidisciplinary and interdisciplinary approaches and made it into a suitable case for several courses, across several disciplines. The proximity gave our students physical access to the places, events, and stakeholders involved, and allowed their involvement along the lines of place-based and case study pedagogies. The students were able to interact with stakeholders, to observe or engage in community discussions and actions, to visit sites along the proposed MVP path, and to monitor policy positions and stakeholder responses in real time. Whether discipline-specific or interdisciplinary, our pedagogy across the courses intentionally teased out civic engagement, individual responsibility, structures of participation, and self-efficacy in environmental issues, in support of the program-level learning goal that students "actively investigate roles and identities as citizens, consumers, and environmental actors."

Two theoretical perspectives seemed particularly relevant for our curricular efforts. First, the "resource curse" model outlined by Sachs and Warner [13] and Partridge et al. [11] helped relate the local pipeline issue to broader regional and global concerns. This model suggests that areas rich in resources often have reduced economic growth due to highly specific and disruptive development that leaves localities dependent on a single resource or industry but does not improve the overall economy of the region and is subjected to revenue volatility (due to fluctuations in the resource market) and lack of economic diversification. Resource-rich areas also often function with diminished democratic systems, because state and local politics tend to be dominated by industry lobbies, which restrict decisions about community development to a small minority and skew redistributive processes [e.g., 16]. Lack of transparency and perceived inequitable power-sharing in resource-rich areas have historically led to conflict and violence [e.g., 6].

Second, the "Sacrifice Zone" framework helped to expand our focus and connect the local pipeline's environmental impact with broader trends in resource development. This model points out that extensive resource development often results in areas that are too degraded to be restored, and are thus considered by developers as an acceptable loss—a "sacrifice zone"—for the resources gained. Permanent degradation is thus thought to be an acceptable aspect of resource extraction. Furthermore, these negative impacts are usually not experienced by residents equally and often "reside in semi-industrial areas—largely populated by African-Americans, Latinos, Native Americans, and low-income whites—where a dangerous and sometimes lethal brand of economic and racial discrimination persists" [9]. What seems typical of the resource-impacted areas examined in these studies held true for the Mountain Valley Pipeline region: that residents "do not have prior experience as activists and do not see themselves as environmentalists" [9].

While the faculty in the program acknowledged the value of these frameworks, there was no expectation that individual courses would address the pipeline within the frameworks. In fact, the only intentional commonality across the courses and co-curricular programming was to use the MVP case to support the ENVS learning goal of student citizenship and engagement. As we discuss below, each faculty member related the MVP case to course-specific themes, theories, and goals, mostly taught within disciplinary frameworks.

2.3.2 MVP Case Implementation in Specific Courses

Across the courses, the MVP case received different amounts of time, ranging from one lab or class discussion to serving as the focus of semester-long student projects. Furthermore, each instructor used the MVP case to illustrate and question discipline-specific concepts and theories. The MVP-related pedagogies relied primarily on three features of the case: access, stakeholder interaction, and intersections with the students' own physical and constructed space, ultimately supporting several of the ENVS program-level learning goals (listed below), as well as course-specific goals. The common thread across all the exercises and co-curricular programs was the opportunity for students to *actively investigate roles and identities as citizens, consumers, and environmental actors* (Program learning goal #1).

Introduction to Environmental Science (ENSC 101): One of the labs in this course consisted of a viewshed analysis along the Appalachian Trail in which students learned a technique for quantifying the aesthetic impacts of the MVP. The viewshed analysis was a quantitative method developed by the National Park Service; one of the MVP stakeholders, the Natural Resource Specialist from the Appalachian Trail Conservancy, led the data collection. The goal of this assignment was to consider how subjective, "beauty is in the eye of the beholder" impacts could be quantified in a standardized way.

Introduction to Environment and Culture (ENST 103): This class centered on a multi-week project that culminated in individual papers and group presentations that addressed interactions with the various MVP stakeholders. Students were assigned to small groups and tasked with interviewing a variety of stakeholders, including landowners, members of Boards of Supervisors (county government), members of the local Chambers of Commerce, and leaders of the protest movement. The interviews were recorded, transcribed, and analyzed for ideological and ethical attitudes, using a chapter on environmental ideology from Julia Corbett's *Communicating Nature*. Additionally, one of the stakeholders, an environmental activist and leader of the local anti-MVP efforts, was invited to the classroom as a guest speaker, with ample opportunity for the students to ask questions and engage in conversation. The assignment served course-level goals of increasing student awareness of environmental ideologies and the primary ethical formulations embedded in citizens' attitudes, bringing environmental theory (ethics and ideology) into the real world of lived experience, and correlating expressions of environmental attitudes with stakeholder relationships to the pipeline.

Introduction to Environment and Society (ENST 105): ENST 105 incorporated the MVP in three separate but connected exercises. The first involved reading local and regional news, exploring the websites of various environmental groups, field observation of community actions, and meeting with stakeholders both in the classroom, as guest speakers, and outside class, at community forums and site visits. The MVP was one of several local current events at the intersection of environment and society to which the students were introduced, alongside sustainable farming, food insecurity, and conservation. This assignment was intended to familiarize students with issues in our area, model community engagement, and encourage them to reflect on their potential for civic effectiveness and on their own interest within the broad realm of "the environment." Second, the local environmental issues explored during the first exercise were used to apply the theoretical frameworks covered during the semester. Through worksheets, group discussions, and group reports to the class, the students considered how, for instance, the political economy perspective might explain a certain aspect of the MVP (e.g., its trajectory; the institutional support it received) differently from the social construction or markets perspective. These applications allowed the students to practice social scientific analysis of environmental issues and to explore new ways for understanding an issue in which the student was becoming an expert. The exercises also served skill-related goals, as they promoted listening and peer discussion and were opportunities for low-stakes oral presentations. Finally, the MVP was the topic of a final "Do Something!" workshop that produced a call to action resolution co-authored by the entire class, suggesting actionable (here, now, by us) solutions to the issue, based on the research and discussions throughout the semester. This assignment, too, served several course-level goals. Most importantly, it was intended to help students realize their potential for citizenship and civic efficacy, reflect on the aspects of environmental issues that prompt them to action, and articulate the role of society in environmental issues.

Critical Zone Science and Management (ENSC 309): Students spent two class sessions evaluating the Environmental Impact Statement (EIS) for the MVP. The class was organized into teams tasked with evaluating different sections of the EIS. Teams then regrouped to discuss the EIS process, the methods used to assess environmental impacts, and potential sources of uncertainty. A subsequent discussion explored the role of science and citizens in the EIS and the FERC review processes and the importance of scientific literacy in informing public discourse. Since this was an upper-level course, the activity assumed that students would already be familiar with the MVP case, allowing for greater focus on course-specific learning outcomes.

Energy Transitions (INQ 271): Students engaged in a multi-week project that culminated in a formal presentation during a class "forum," with each student assigned to represent an MVP stakeholder. Unconventional gas transmission pipelines are matters of present concern; students were asked to apply historical understanding and use research skills to frame a coherent position. They were required to sift and winnow many sources of information and consider multiple perspectives to clarify their (and our) thinking about the MVP project. Based on their preferences, the students divided into small groups or worked individually to investigate a particular stakeholder. The professor provided some initial sources to serve as a starting point for the students' research and in-class coaching to direct their research in the weeks leading to the presentation, which took place toward the end of the class. Within the scope of the course, the purpose of the forum was to practice effective public speaking (oral presentation) skills, as well as to deepen student understanding of the historical forces driving natural gas pipeline development by considering a case study in our "backyard," the MVP.

Geographic Information Systems (ENSC 270). GIS students used the pipeline case study as an illustration of several types of spatial analysis employed in public discussions about the MVP. In one activity, students were provided with published maps of the pipeline produced by EQT, the *Roanoke Times*, and opponents of the MVP and asked to consider the idea of maps as models of reality and the ways in which map design can influence interpretation. The class also conducted a viewshed analysis using GIS that built on the field experience in the introductory class and reverse-engineered a widely disseminated photorealistic visualization of the pipeline. As with the Critical Zone Class, the assumption was that students were already familiar with the MVP case, allowing for greater focus on course learning objectives.

2.3.3 Co-curricular Events Related to the MVP

In the year prior to implementing the MVP case, our program began engaging with the pipeline and its stakeholders. In Fall 2016, the Environmental Studies Program co-sponsored a public forum, in collaboration with the Roanoke Appalachian Trail Club, Roanoke Valley Cool Cities Coalition, and the Blue Ridge Land Conservancy. The event included a wide range of perspectives from experts and stakeholder group representatives on both sides of the issue. For instance, the President of the Roanoke Regional Chamber of Commerce spoke in favor of the MVP's construction for its potential economic benefits; Appalachian Mountain Advocates opposed the MVP because of its expected negative economic impacts in the region; and experts including a geologist and a hydrologist, as well as the Appalachian Trail Conservancy, spoke against the pipeline because of its negative environmental impacts. FERC and EQT (the company developing the MVP) were not represented at the forum, despite having been invited. The forum was a co-curricular event intended to engage ENVS students in considering multiple perspectives on this controversial pipeline, which was then in the planning and permitting stage. Some of the students prepared posters of their MVP-related projects conducted in ENVS classes. Given the interests and expertise of the speakers, the forum also highlighted a variety of methodologies from the sciences and social sciences for analyzing the issue, as well as various environmental ethics.

The following Fall (2017), ENVS co-sponsored the Forum for Health Professionals: Health Implications of Energy Choices in SW Virginia, in collaboration with Physicians for Social Responsibility and the Greater Roanoke Valley Air Quality and Asthma Coalition at the Virginia Tech's Carilion Medical School. The MVP was featured in several presentations during the forum, including a talk by a psychologist on the mental health impacts of the legal wrangling in conjunction with the pipeline and the impacts of worry about water cleanliness and climate change; a talk by a hydrologist on water safety; and a talk by a geologist on the risk of landslides, explosions, and direct injury during and after the construction of the pipeline. For the ENVS Program, this served as a co-curricular event that encouraged our students to consider the health and wellness impacts of the planned pipeline.

2.3.4 Course-Level Assessment

We assessed the program-level goal of having students "actively investigate roles and identities as citizens, consumers, and environmental actors" across all the courses that incorporated the MVP case in Fall 2017, with a survey instrument comprised of 11 questionnaire items. The first four items measured the students' perceived efficacy, civic engagement, issue involvement, and environmental issue awareness; items 5–11 collected demographic data, such as gender (female/male/non-conforming), race/ethnicity, and religion (Table 2.1). Two of the questionnaire items (Item #2 and Item #5) were multiple choice, the rest were open-ended.

We administered the questionnaire twice, once in the first week of classes and again in the last week of Fall 2017, in a pseudo-experimental design. The course material incorporating MVP-related activities throughout the semester constituted the "treatment." Because this research design does not include a control group, we are not able to separate the effects of the MVP-related activities from other course material, nor can we measure the effect of ENVS courses relative to other courses at Roanoke College. As such, we cannot claim generalizability or causation in our findings. However, the data do allow us to gauge whether there was a difference in the students' perceptions of their roles as citizens, consumers, and environmental actors after participating in courses that incorporated case-based and place-based applications throughout the semester.

Table 2.1 Course-level assessment tool

Item #	Item wording
1	To what extent do your actions as a citizen and consumer make a difference? Please explain.
2	Rate your interest in making a difference in the world on a scale from 1 (least interest) to 5 (most interest).
3	On what issues do you want to make a difference?
4	Name some current environmental issues.
5	Gender (circle one): female; male; non-conforming
6	Race/ethnicity:
7	Religion:
8	Major:
9	Class year:
10	What is your hometown? City: State: Country:
11	Extracurricular activities:

2.4 Assessment: How Successful Was the Incorporation of the MVP Case?

2.4.1 Program-Level

As noted earlier, although the primary focus of this initiative was on the use of the MVP case at the program-level, each instructor explored the case from their own disciplinary perspective, allowing for in-depth treatment of the MVP through the lens of the natural sciences, humanities, and social sciences. The intention was that students enrolled in several of these courses simultaneously would be exposed to multiple disciplinary perspectives. The instructors further facilitated interdisciplinary connections through discussion and questions prompting students to reference MVP material from their other courses, albeit to varying degrees. Co-curricular events provided additional opportunities for interdisciplinarity by allowing students to witness first-hand a range of stakeholder viewpoints rooted within different frames of nature, economics, justice, history, and ethics. These shared co-curricular events also allowed students from across all areas of the program to engage in conversations with one another and to exchange discipline-specific knowledge and viewpoints.

The MVP case integration had several positive externalities for the ENVS program as a whole. From a faculty perspective, the program-wide initiative led to more frequent and more meaningful interactions in which we exchanged ideas about assignments and resources, discussed what worked and what did not in class-specific student activities, and ultimately co-authored a conference presentation and this chapter. This shared case study reinforced the culture of collaboration among faculty in the program. Additionally, faculty and student participation in MVP-related events deepened and expanded the program's connections with community stakeholders. The program was able to provide support for some community initiatives, while also strengthening relationships with some of our partners, who became valuable resources for our students and for our teaching.

In short, the MVP case, as well the willingness of ENVS faculty to revise their course designs to incorporate this case, provided a rare opportunity to be intentional about implementing interdisciplinarity at the program level, something very difficult to accomplish otherwise. Given its features, the case also presented an opportunity to pursue and evaluate program learning goal #1—that students *actively investigate roles and identities as citizens, consumers, and environmental actors*—across the six courses. From an implementation perspective, we argue that this pedagogy was a success and we discuss in a later section the lessons learned from these integrative and interdisciplinary approaches. However, from the perspective of program learning goal #1, on which we focused our student assessment, the results are, at best, inconclusive.

2.4.2 Survey Results

The number of valid observations in our sample was 100. One hundred twenty-one students enrolled in the six participating courses—ENSC 101, ENST 103, ESNT 105, ENSC 270, ENSC 309, and INQ 271—were surveyed in the first week of Fall 2017; after we implemented the survey again, in the same courses, at the end of the semester, we collected 100 usable responses. This reduction in observations happened for several reasons: some students dropped out of the sample, others did not complete the questionnaire at the end of the semester, yet others were enrolled in several of the courses and we only used one of their response sets. At the data collection stage, students were assigned random numbers, which they used when they submitted their surveys; these numbers were employed by the research team to link the pre- and post-treatment responses. In terms of demographics, our sample was majority white, reflective of the demographics of the region and the college; majority female, with approximately 60% of the respondents identifying as such; divided equally between ENVS and other majors, including undeclared students; and inclusive of all years in college but representing first years and juniors more heavily than sophomores and seniors. Based on their stated hometowns, a majority of the respondents are from areas proximate to the MVP pathway and to the Atlantic Coast Pipeline (Fig. 2.2).

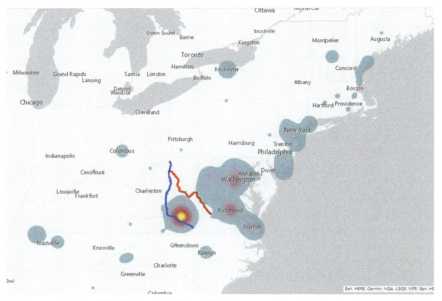

Pipeline centerlines from DPMC_GIS (hosted on ESRI ArcGIS On-Line; accessed 6/14/2018) based on data provided to FERC.

Fig. 2.2 Geographic distribution of respondents, based on self-identified hometown

For this analysis, student assessment data was aggregated across all participating courses, to gauge the effectiveness of the MVP case for teaching responsible and engaged citizenship at the program level, rather than in individual courses. In this section, we focus on the data from questionnaire items #2, #3, and #4 (Table 2.1): the students' interest in making a difference in the world; issues on which the students may want to make a difference; and environmental issues of which the students are aware. We exclude from discussion question #1, intended to measure the respondents' perceived civic efficacy by asking, "To what extent do your actions as a citizen and consumer make a difference? Please explain." A quick glance at the data collected on this question suggests that it was ambiguously worded and did not measure what we intended it to measure. Instead of addressing perceptions of their own ability to make a difference in their community or the world, the majority of the respondents provided examples of ways in which "one" may influence the environment (e.g., recycle, use public transportation). Questions 2, 3, and 4, on the other hand, have relatively high face validity. Since the research design was pseudo-experimental, we refer to the beginning of semester measurement as pre-treatment and the end of semester measurements as post-treatment; the material in the participating courses and co-curricular events is equivalent to the treatment.

On the **engagement question** (questionnaire item #2), students rated their interest in making a difference in the world on a scale from 1 (least interest) to 5 (most interest). The median value for all respondents was 4 in the pre-treatment survey and remained unchanged in the post-treatment survey. In this sample, the respondents started with relatively high interest in making a difference in the world and, overall, appeared to maintain the same level of interest. However, a closer look at the distribution of cases across answer choices hints to slight shifts toward more interest. In the pre-treatment survey, 25% of the respondents rated their interest in the low (1–3) categories, compared to only 17.4% post-treatment; 34.4% of the respondents offered an answer of 4 (high) pre-treatment and 43.3% answered 4 post-treatment; 40.6% answered 5 (very high) pre-treatment and 39.2 provided the same answer post-treatment. Although we do not evaluate statistical significance, the data suggest that interest in making a difference in the world increased somewhat during Fall 2017, with the 7.6% decrease in categories 1–3 (low interest) and the 8.9% increase in respondents self-selecting in category 4 (high interest) post-treatment.

In terms of **issues on which the students wanted to make a difference**, questionnaire item #3, there was relatively little agreement pre- or post-treatment. Although pollution, climate change, waste, energy, and food received the most attention, respondents reported interest in a wide range of issues including deforestation, poverty, ocean pollution and acidification, habitat loss, animal welfare and abuse, food waste on campus, homelessness, hunger, health of returning veterans, gender equality, racial injustice, environmental disparities, lack of support for science, lack of knowledge of farming, women in farming, bees and pollinator health, overpopulation of humans and over-consumption of resources, and more. There is no marked difference in focus on certain issues at the beginning and the end of the semester.

The MVP did not feature highly as an issue on which students wanted to make a difference neither at the beginning of Fall 2017, nor at the end of the semester. In the pre-treatment survey, only one respondent mentioned "Mountain Valley Pipeline" as an issue on which they wanted to make a difference. One mentioned fracking and one mentioned "local issues—especially pipeline." All three listed hometowns in Virginia. In the post-treatment survey, three respondents mentioned the pipelines, with two mentioning MVP by name as an issue on which they wished to make a difference. All three were second-year students: two from Virginia, one from New Jersey; two women and one man (self-reported).

In terms of **environmental issues of which students were aware**, questionnaire item #4, climate change was the most named issue in the pre-treatment survey. Climate change remained an important issue in the post-treatment survey as well, but students also mentioned pollution, waste, ocean, energy, and food with relatively high frequency. "Pipelines" as an environmental issue also gained some visibility in this group of students after participating in the curricular and co-curricular ENVS programming in Fall 2017. Pre-treatment, 3% of the respondents named the local MVP or "Virginia pipelines" as current environmental issues, with a total of 12% naming fracking or pipelines more broadly, including oil pipelines as well as gas transmission lines. Post-treatment, 11% of students named the MVP as an environmental issue; another 10% named pipelines and an additional 5% listed "fracking natural gas" but not pipelines. Overall, 21% of the post-treatment respondents listed pipelines (26% if we also include the mentions of fracking) among current environmental issues, a 14% increase compared to the pre-treatment results.

Overall, it appears that the MVP teaching units did raise awareness of pipelines as environmental issues. There is also some evidence that the students who completed the six courses that were part of the MVP integrative initiative registered a slight overall increase in their interest in civic engagement, measured as their interest in making a difference in the world. However, given the design of our assessment tool and of the study, we cannot make any conclusive causal claims regarding the effects of MVP-related activities on the students' engagement as citizens and environmental actors or on their (re)definition of community and sense of place in this area.

2.5 Lessons Learned

Environmental Studies curricula are often structurally challenged by the need to address multiple content areas, competing theories of social change, and shifting political landscapes. In the absence of a central set of organizing theories, this "big tent" approach to Environmental Studies may contribute to a sense of what Torsten Husen [as cited in 15] referred to as "multi-disciplinary illiteracy" in which students are sent forth with a broad but shallow understanding of many disciplines combined with a limited sense of how these disparate pieces combine into a coherent whole. More recently, Environmental Studies educators have called for greater integration of theory into ES curricula [12], the development of a central canon [15], explicit discussion of theories of social change and conflict resolution [10], and a transition of ES programs from an emphasis on multidisciplinarity toward a more cohesive state of interdisciplinarity, which can only be achieved "when there is sustained interaction on a formal and informal basis between members of different disciplines" [2, 15]. Case studies that explore complex "wicked" problems and provide both students and faculty with a focal point for exploring connections between environmental theory and practice may, thus, provide an important first step toward helping prepare ES students to become "thoughtful and anticipatory agents of change in the tumult to come" [10].

Within our curriculum, the use of a singular MVP case in multiple courses provided an organizational focus for exploring a set of complex, local issues from different disciplinary perspectives, with varied degrees of depth, and through a wide range of activities. Importantly, the case had three salient features that made possible these integrative program-level pedagogies. The first was the pipeline's physical proximity to the students, which facilitated access and allowed students to "experience" the MVP case while also learning in an adjacent space. The second feature was the opportunity for stakeholder interaction and community engagement, as many and diverse individuals and groups in the surrounding community contributed to an ongoing conversation, engaged in actions intended to modify behavior and policy, and were open to interactions with our students. The third was the presence of the MVP in a space adjacent to or intersecting the students' own constructed space, allowing them to develop empathy and explore their role as citizens as they engaged with an issue that affected them, their community, and their place.

2.5.1 Crossover of Pedagogical Approaches

While the ENVS faculty agreed on the value of incorporating the MVP within their individual courses, there was no centralization of the course-specific MVP-related pedagogical choices. The only intentional commonality across the courses and co-curricular programming was to use the MVP case to support the ENVS program learning goal that students "actively investigate roles and identities as citizens, consumers, and environmental actors." It is interesting, then, that the same features of the case constituted the basis for activities in courses across different disciplines. For instance, the physical proximity of the MVP led to the viewshed analysis exercise in ENSC 101, a natural science course, and field observation of community action in ENST 105, a social science course. These types of activities made possible by direct access align with the literature on case study pedagogies and place-based learning, as they are experiential and grounded in the real world [14] and incorporate both teachings by example (i.e., building knowledge from inductive rather than deductive reasoning) [1] and active learning to promote translation of theory into practice [19].

2.5.2 Complexity

The Mountain Valley Pipeline crosses municipalities and rural counties marked by class differences and divided political affiliations; students at Roanoke College also represent a range of socioeconomic privilege, political beliefs, and place attachments. As such, engaging with the MVP across the curriculum required foregrounding socioeconomic difference while still retaining a focus on the key concepts and theories taught within disciplinary frameworks. Although the underlying complexity is a key reason why the MVP represented such a rich opportunity as an integrating case study, the overlapping legal and regulatory points, economic impact forecasts, environmental impact assessments, and environmental justice dimensions may also have presented barriers for students to fully engage with the issues. These academic and technical complexities were further compounded by the emotional charge, implicit biases, and overwhelming sense of powerlessness associated with the environmental justice dimensions of the case.

Adding to the level of pedagogical challenge, the case was evolving as we were teaching it. For some courses, the dynamic nature of the case presented opportunities to examine shifts in policy in real time. Yet for courses focused on other aspects of the case, the shifting nature of the discussion challenged students' ability to contextualize and synthesize information, particularly in introductory courses. On a more practical level, the "happening now" and "in the news" nature of the case increased the difficulty of planning activities, field trips, and events in ways that reflected the most recent developments.

Along the same lines, the opportunity to engage with stakeholders representing diverse viewpoints and areas of expertise was a pedagogical asset, but may also have contributed to a sense that the problem was overwhelming, intractable, and difficult to comprehend. Each stakeholder emphasized different aspects of the issue and offered their interpretation. While, in reality, there was significant overlap in stakeholder positions and broad areas of agreement existed, time and resource constraints allowed only for fragmented glimpses of stakeholder viewpoints, which did not clearly sum up to overlap and consensus. The challenge of framing a cohesive story within a complex and shifting landscape reflects what Soule and Press [15] referred to in stating, "Students already must cope with the complex nature of environmental problems; now, in addition, they are confronted by a spectrum of ideologies that promote conflicting problem definitions, analyses, and favored solutions."

2.5.3 Proximity

Roanoke College is located in close proximity to the MVP, but not immediately affected by the construction of the pipeline. This physical proximity provided our students with access to the places, events, and stakeholders involved, and allowed their engagement along the lines of place-based and case study pedagogies. Students were able to interact with stakeholders and to observe or engage in community discussions and actions. Yet, in a number of ways, the MVP's proximity, which we initially viewed as an opportunity to broaden student definitions of community and redefine their sense of belonging, represented both an advantage and a disadvantage as a case study. For some students, the physical proximity increased their engagement, empathy, and interactions, allowing them to form deeper connections with both people and spaces that would be affected by the MVP and to revisit their conceptions of place and community. For others, the peripheral location of the pipeline to their own constructed space, both physically and conceptually, may have proved a barrier to engagement. They may have acknowledged the presence of the MVP nearby, but did not revise their definitions of their own place, community, and self to incorporate the pipeline and its effects. In the end, while some students did engage in the MVP-affected communities and showed empathy, others did not. We speculatively attribute these differences in student reactions to personality, socialization, integration, and identity, which influenced the MVP's position relative to their own universe.

2.5.4 Stakeholder, Community, and Student Engagement

Stakeholder interactions were incorporated into most courses, albeit in different ways, including: instructor-facilitated stakeholder guest presentations in ENST 103 (a humanities course) and ENST 105 (a social science course); student-directed stakeholder interaction in ENST 103, where the students were tasked to interview landowners, Chambers of Commerce, and County Boards of Supervisors; stakeholder-directed tasks in ENSC 101, where the Natural Resource Specialist from the Appalachian Trail Conservancy led the viewshed analysis lab; and the Fall 2016 co-curricular stakeholder forum. Regardless of discipline, the instructors acknowledge that such interactions helped to promote connections to the community [14], critical thinking skills, knowledge acquisition, and an enhanced ability to make connections and view issues from different perspectives [7, 20].

While participating faculty agreed that implementation of the MVP case provided numerous benefits to ENVS faculty and students, in hindsight, we also recognize that we may have overlooked important opportunities for our program to give back to the community. Some of the challenges imposed by logistics and the shifting nature of the case may have been eased through greater involvement of community partners during the development phase of our pedagogical project and greater coordination throughout the semester. Student engagement in the pipeline outside the classroom may also have benefitted from a more coordinated effort on the part of faculty to facilitate and even model ongoing engagement with community partners, perhaps through a culminating experience that provided students with a concrete way to integrate their classrooms experiences and contribute to the community. This culminating experience may also have helped students to gain a sense of agency and empowerment.

2.5.5 Interdisciplinarity at the Program-Level

We conceived of the MVP case as an interdisciplinary opportunity at the program-level rather than at the level of individual courses with the intention that students enrolled in multiple courses would have an opportunity to experience the case through multiple disciplinary lenses. However, the reality was that only a small number of students were enrolled in multiple participating courses. For those who only took one of the MVP classes, the guest speakers and co-curricular events may have been insufficient for shifting their analysis of the subject toward the interdisciplinary realm. We also limited the coordinated integration of the MVP across the curriculum to a single semester and the majority of the students who learned with the MVP case in Fall 2017 were first-year or sophomore-level students.

The program was not intentional in offering an opportunity for the students to follow up on the MVP case, work on solutions, build on their knowledge in subsequent and more advanced courses, or take classes later on that addressed the MVP from a different disciplinary perspective. We likely missed an opportunity to capitalize on the complexity and interdisciplinarity of the case, as we did not maintain focus on it for additional semesters or returned to it in an integrative, program-level way in the subsequent academic year. Anecdotal evidence based on the courses that used the MVP case, independently, in Fall 2016 suggests that students in those courses carried some interest in and knowledge of the case into their Fall 2017 courses. This was most evident in the upper-level courses, ENSC 270 and ENSC 309, where students were able to comment on the MVP and use it in applications because they were already familiar with it from courses in the previous year.

2.5.6 Assessment and Research Design

Integrating the same case study across multiple disciplines and courses was ambitious and, understandably, the majority of our energy was focused on time-sensitive aspects of course management such as developing new materials, organizing field trips, integrating stakeholders in classes without exhausting them, and ensuring stakeholder responsiveness when students were assigned interviews or projects. The addition of an assessment tool was proposed relatively late in the process, with the added pressure of obtaining IRB approval in time to implement the survey during the first week of Fall 2017. Development of a sound research design was not a significant part of our conversations during the planning stages, either, and it was not until we started to work on the co-authored pedagogical project that we discussed alternatives for designing a study with higher internal and external validity (e.g., establishing control groups).

As a result, there was little time for searching the literature for pre-tested, similar surveys and experimental designs, or running a pilot to evaluate the quality of our questionnaire items. In the end, most questions had high reliability and validity, but measures of students' perceived civic efficacy (item #1), central to the program goal we were assessing, fell short. Furthermore, there was no expectation or coordination on how each instructor would incorporate the MVP in their courses and that discipline-specific concepts and learning objectives varied by course. The measures we designed had to be rather generic. Civic efficacy and environmental justice are both value-laden and complex concepts. More nuanced measures that contextualized engagement in the course material and language would have produced richer results.

2.6 Summary

Integrating the MVP across ENVS curricular and co-curricular programming in Fall 2017 was exhilarating, enriching, overwhelming, and exhausting. While our design and implementation of this interdisciplinary project had challenges, we want to recognize the benefits as well. For the students, this was an opportunity to interact in real time with an issue that affected many in the communities surrounding Roanoke College and to experience first-hand the complexity of environmental issues, which renders interdisciplinary analysis and problem-solving essential. For our program, this project led to community-building and deeper collaboration, while also pushing us to articulate what interdisciplinarity means and how it can be taught in the context of Environmental Studies at Roanoke College.

Would we do it again? In a heartbeat. Would we do it differently? Somewhat. We would keep the various applications developed for the courses, the activities that capitalized on the physical proximity on the MVP, and the engagement with the active network of community stakeholders involved in the issue. We would also continue with the interdisciplinary pedagogical conversation we started. In terms of changes, we would increase the depth of coordination of course-level goals, with more discussion of theoretical and conceptual integration from various disciplines; we would increase the involvement of community stakeholders at the planning stages, including a more purposeful exploration of ways to give back meaningfully and productively to the community; we would follow through in courses over several semesters, for knowledge enforcement, deeper understanding, and more opportunities for interdisciplinarity, including a practical, solution-focused opportunity for the students at a later point in the program. From the point of view of contributions to pedagogical scholarship, we would develop the assessment tool concomitantly with pedagogical development.

References

1. Allchin D (2013) Problem-and case-based learning in science: an introduction to distinctions, values, and outcomes. CBE—Life Sci Educ 12(3):364–372
2. Braddock RD, Fine J, Rickson R (1994) Environmental studies: managing the disciplinary divide. Environ 14(1):35–46
3. Burns W (2019) The case for case studies in the context of environmental issues—updated. Case Stud Environ 3(1):1–5
4. Clark SG, Rutherford MB, Auer MR, Cherney DN, Wallace RL, Mattson DJ, Clark D, Foote L, Krogman N, Wilshusen P, Steelman T (2011) College and university environmental programs as a policy problem (part 1): integrating knowledge, education, and action for a better world? Environ Manage 47(5):701–715
5. Cronon W (1995) Uncommon ground: toward reinventing nature. WW Norton, New York, p 65
6. Le Billon P (2005) Fueling war: natural resources and armed conflict. Routledge, London
7. Lee K (2007) Online collaborative case study learning. J CollE Read Learn 37(2):82–100
8. Lekan T (2014) Toward a problem-centered approach to environmental studies: challenges and prospects. RCC Perspectives 2:37–42
9. Lerner S (2010) Sacrifice zones: the front lines of toxic chemical exposure in the United States. MIT Press, Cambridge
10. Maniates M (2013) Teaching for turbulence. In: Institute Worldwatch (ed) State of the world. Island Press, Washington, DC, pp 255–268
11. Partridge M, Betz M, Lobao L (2012) Natural resource curse and poverty in Appalachian America. MPRA Paper 38290, University Library of Munich, Germany
12. Proctor JD, Clark SG, Smith KK, Wallace RL (2013) A manifesto for theory in environmental studies and sciences. J Environ Stud Sci 3(3):331–337
13. Sachs JD, Warner AM (1999) The big rush, natural resource booms and growth. J Dev Econ 59:43–76
14. Sobel D (2004) Place-based education: connecting classrooms and communities. Orion Society
15. Soule ME, Press D (1998) What is environmental studies? Biosci 48(5):397–405
16. Thomson V (2017) Climate of capitulation: an insider's account of state power in coal nation. MIT Press, Cambridge
17. Vincent S, Focht W (2011) Interdisciplinary environmental education: elements of field identity and curriculum design. J Environ Stud Sci 1(1):14–35
18. Vincent S, Bunn S, Stevens S (2012) Interdisciplinary environmental and sustainability education: results from the 2012 Census of U.S. Four Year Colleges and Universities Council of Environmental Deans and Directors
19. Wei CA, Burnside WR, Che-Castaldo JP (2015) Teaching socio-environmental synthesis with the case studies approach. J Environ Stud Sci 5(1):42–49
20. Yadav A, Lundeberg M, DeSchryver M, Dirkin K, Schiller NA, Maier K, Herreid CF (2007) Teaching science with case studies: a national survey of faculty perceptions of the benefits and challenges of using cases. J CollE Sci Teach 37(1):34

Chapter 3
Learning to Undermine a Pipeline: A Multi-logue on Encounters with Vermont's Addison Natural Gas Project

Julie Macuga, Ingrid L. Nelson, Rachel Smolker, Trish O'Kane, and Brian Tokar

Abstract In 2013, Vermont Gas Systems (VGS) received a certificate of public good for their proposed 41-mile pipeline extension called the Addison Natural Gas Project (ANGP) for transporting fracked gas from Alberta, Canada via the TransCanada Mainline. The ANGP is the largest expansion of fossil fuel infrastructure in Vermont in decades. Activists, those working and studying in higher education, local residents, and others have resisted the ANGP with diverse strategies. An investigation launched in 2017 regarding the pipeline's ongoing safety and construction problems remains open. In co-writing this chapter, we offer brief autobiographical comments, followed by a "multi-logue" consisting of a facilitated and recorded discussion, which we transcribed and circulated among ourselves for revision and reflection. Our "multi-logue" writing experiment highlights our multiple perspectives on the ANGP and on broader pipeline pedagogies and entanglements. We reflect on our experiences learning to undermine the ANGP through two themes: (1) strengthening human and more-than-human relationships with places threatened by pipeline expansion, and (2) practicing pipeline pedagogies amidst Higher Education Institution (HEI) politics.

J. Macuga
350Vermont, Burlington, VT, USA
e-mail: resist@350vt.org

I. L. Nelson (✉)
Department of Geography, University of Vermont, Burlington, VT, USA
e-mail: ilnelson@uvm.edu

R. Smolker
Biofuelwatch and Protect Geprags Park, Hinesburg, VT, USA
e-mail: rsmolker@riseup.net

T. O'Kane · B. Tokar
Rubenstein School of Environment and Natural Resources, University of Vermont, Burlington, VT, USA
e-mail: Patricia.OKane@uvm.edu

B. Tokar
e-mail: Brian.Tokar@uvm.edu

Keywords Addison natural gas project · Multi-Logue facilitated discussion · Vermont · Higher education institution politics · Pipeline pedagogy

3.1 Introduction

Our sense of place thrives with the people, birds, trees, soils, and others in our midst. In the State of Vermont, we live and work on Abenaki lands and grapple with the enduring legacy of settler colonialism and Vermont's eugenics interventions in the 1920s and 1930s [see 21, 7]. We are struggling for more just and climate-conscious communities as we work to connect spaces of learning and resistance. And so, we regret the necessity of beginning with details about an unwelcome recent arrival in our midst: the Addison Natural Gas Project (ANGP) pipeline.

3.1.1 The ANGP

In December of 2013, Vermont Gas Systems (VGS) received a certificate of public good for their proposed 41-mile pipeline extension transporting fracked gas from Alberta, Canada via the massive TransCanada Mainline. The ANGP is the largest expansion of fossil fuel infrastructure in Vermont in decades (the previous extension ended in Colchester and it supplies homes in the Burlington area). The 2013 extension had three planned phases, with Phase I extending southward to Middlebury. Phase II would have extended under Lake Champlain to Ticonderoga, NY, but was canceled. Phase III has been postponed indefinitely, but would have extended south to Rutland, VT and ultimately the regional Marcellus Shale gas infrastructure.

Activists, those working and studying in higher education, local residents, and others have resisted the ANGP with diverse strategies, developing what geographer Kai Bosworth [3] refers to as populist pipeline opposition movement identities and practices of counter-expertise along the way. Grassroots organizers supported by 350Vermont also successfully campaigned in 2019 to cancel fracked gas infrastructure buildout in Bristol, VT. An investigation launched in 2017 regarding the pipeline's ongoing safety and construction problems remains open (see [13] for background and [20], for key investigation documents).

3.1.2 A Multi-logue Writing Approach

In co-writing this chapter, we offer different perspectives on the ANGP and on broader pipeline pedagogies and entanglements using a "multi-logue" format consisting of a facilitated and recorded discussion, which we transcribed and circulated among ourselves for revision and reflection. We draw inspiration from Larissa Barbosa da Costa and colleagues' [5, p. 260] "trialogue" approach to thinking and writing together, highlighting points of convergence and friction amidst our personal experiences and collective thinking. In contrast with da Costa and colleague's [5] context of knowing one another for four years, some of us have worked together for many years while some of us met for the first time for this project. We have complicated feelings regarding academia, as we hold and have previously held different positions within and outside of academic spaces at different times in our lives.

In our multi-logue, we reflect on our experiences learning to undermine the ANGP pipeline through two themes: (1) strengthening human and more-than-human relationships with places threatened by pipeline expansion, and (2) practicing pipeline pedagogies amidst Higher Education Institution (HEI) politics. Although separated in the text, these themes overlap and highlight occasional contradictions. Writing together in a multi-logue format presents challenges regarding structure, flow, and the prominence of some voices over others. We certainly have not resolved all of these issues, but we hope this experiment offers one example of many possibilities for writing pipeline pedagogies together. Perhaps this multi-logue reflects what "staying with the trouble" might look like in times of climate crisis [10]. We are thankful for the opportunity to share our thoughts and questions regarding pipeline pedagogy based on our grounded experiences. Before doing so, we introduce ourselves in our own words.

3.1.3 Auto-Biographical Comments

Julie Macuga (J): I am 350Vermont's Extreme Energy Field Organizer, a graduate of the University of Vermont (UVM), and an underminer of the fossil fuel economy. My work as an activist began in 2016 when I took co-author Trish O'Kane's "Birding to Change the World" class at UVM. I learned about local residents and activists defending Geprags Park from the pipeline extension. The park is home to rare songbirds called golden-winged warblers, which became a prominent symbol of ANGP resistance. After participating in a non-violent direct action by putting my body on the line to prevent the pipeline's construction, I began volunteering with the grassroots group Protect Geprags Park,[1] and researching the companies behind the pipeline obsessively. Eventually, I bought shares in Energir, VGS' parent company, and traveled to Montreal to attend (and disrupt) their annual meetings in order to give voice to what was happening in Vermont. This led to my current position with 350Vermont—where I organize local and regional resistance to the fossil fuel industrial complex in the many forms that takes.[2]

Maeve McBride (M): I am the former director of 350Vermont, a statewide independent affiliate of 350.org, a global organization building a movement for climate

---

[1] https://www.protectgeprags.org/.
[2] https://350vermont.org/extreme-energy/#vtgas.

justice. I was also a core organizer and leader in the Stop the Fracked Gas Pipeline campaign, the multi-year coalition campaign to resist the expansion of fracked natural gas infrastructure in Vermont. In that capacity, I helped set strategy, foster grassroots organizing, organize demonstrations and direct actions, negotiate challenges among groups and organizers, and interface with student activists, student interns, student climate groups, and allied faculty. I have had various roles at UVM including doctoral student, community fellow for the Economics for the Anthropocene graduate program, and adjunct faculty in both the School of Engineering and the Department of Geography.

Rachel Smolker (R): I grew up immersed in nature as the daughter of two biologists. My father was an ornithologist, and founder of the Environmental Defense Fund. As a young adult I studied zoology and traveled extensively doing field studies. It became apparent that studying the creatures I loved was not enough as they were increasingly threatened, so I turned my attention from academics to activism. I have been co-director of an international organization, Biofuelwatch, working on climate and land use, human rights and biodiversity protection since 2007. When the Vermont Gas pipeline construction equipment arrived at the edge of my home- town of Hinesburg, I stepped into the ongoing resistance as a founder of Protect Geprags Park. I've worked on a lot of different kinds of resistance from community organizing to UN negotiations. Like many, I am just doing everything I can possibly manage to stem the tide of destruction that we are witnessing, and hanging on tight to visions of a better world for my children!

Trish O'Kane (T): I teach environmental studies at UVM. Maeve McBride's emphasis on combating the academic colonization of communities is key to my work. In my teaching, writing and citizenship, I strive to decolonize by building long-term relationships. I first learned this as a wide-eyed graduate in the 1980s when I was recruited by radical Jesuit priest and economist—Xabier Gorostiaga, sj—to work at his Sandinista-Freirean research institute in Nicaragua. I keep relearning about relationship-building in the UVM "Birding to Change the World" class I teach, which pairs undergraduates as enviro-mentors with children in local schools.[3] Every Wednesday I take my student flock to Flynn Elementary School to play and learn with neighborhood children and birds for three hours, not as colonizers parachuting in from the Ivory Tower, but as "co-explorers" on a grand adventure, learning from the kids who know their local woods, marshes and lakeshore. Slowing down long enough to get to know a child, a bird, a river, a tree, is a way to resist our frantic, metastasizing capitalist culture and the corporatization of our education system. It can also give us the emotional fuel to fight for systemic change.

Brian Tokar (B): My initial academic training was in mainstream laboratory biophysics and neurobiology, but after a decade away from academia, I came to teach social ecology for ten years at Goddard College in Vermont, which was founded by students of John Dewey in the 1930s. Goddard has preserved a strong focus on

---

[3] See https://www.youtube.com/watch?v=RyzTkooqPXQ.

progressive and experiential education over more than eighty years. When I started teaching at UVM in the mid-2000s, I aimed to bring as much of my progressive education experience as possible into the university classroom. All of my intermediate-level Environmental Studies classes at UVM include a relatively open-ended group project component, where the only fixed requirement—other than for clear reflection and documentation—is that student projects need to have a substantial public component. For several years, participation in and analysis of events around the ANGP were among the most popular project choices. I first learned about the ANGP from Jane and Nate Palmer, the first two landowners to publicly oppose the pipeline. I met them when they were guest presenters in a colleague's agroecology seminar, where I was also a guest panelist. Soon after that, a group of former UVM Environmental Studies students started working with the Palmers to launch a more public effort. They began as a small five-person affinity group, which eventually became the Vermont chapter of the international climate direct action network, Rising Tide.[4] The construction of the full pipeline was once seen as inevitable, supported by a Democratic governor and several statewide environmental groups at the outset. But the dedicated and highly focused organizing by Rising Tide Vermont eventually led to the cancellation and postponement of the pipeline's second and third phases respectively.

Ingrid L. Nelson (I): I began teaching geography and environmental studies at UVM in 2013. I asked student project groups in my geospatial technologies course to study a spatial question in Vermont. One group proposed mapping the planned ANGP with available environmental and social data layers. They earnestly contacted VGS for the ANGP spatial data and were upset when rebuffed, as they knew faculty in UVM's Consulting Archaeology Program who had the proprietary data as part of over $200,000 of sub-contracted impact assessment work. Students eventually obtained a poor-quality paper map distributed during a public consultation meeting and they resorted to digitizing the pipeline from this and other ancillary information. The following year, I co-advised a student writing her senior thesis about critical community organizing pedagogies through her experiences co-teaching a "Students-Teaching-Students" course titled "Community Organizing and Environmental Activism", which combined social movement readings with skills trainings the students learned through attending the Greenpeace[5] Semester in Washington, D.C., the 2013 Power Shift[6] event in Pittsburgh, PA, and from their work in the Vermont Student Climate Coalition (VSCC) [9]. These teaching and mentoring experiences inspired my research on the politics of HEIs 'going green' in the Green Mountain State (Vermont). From data access struggles to critical perspectives on activism and cultures of sustainability expertise, I continue to learn with students such as Julie and many others who are working to undermine pipelines.

---

[4] https://risingtidenorthamerica.org/.
[5] https://www.greenpeace.org/.
[6] https://www.powershift.org/.

3.2 Multi-logue

3.2.1 The Golden-Winged Warbler: Strengthening Place-Based Relationships

On the morning of Thursday, June 6th, we met in the 350Vermont office in downtown Burlington to record our conversation, using the editors' questions about pipeline pedagogies as a reference.[7] Ingrid introduced the idea of the chapter and that perhaps environmental activists, students, professors and others or maybe even those unfamiliar with pipeline resistance might find these kinds of conversations and questions interesting, helpful, informative, controversial or all of the above…

> *I* : I was wondering if we could start with Julie to get a sense of how you got here…not to put you in the spotlight as some sort of superhero, as that's not what this is about. But when thinking about pedagogy we rarely hear in-depth views from [former] students in their own words. Maybe if we can start with your experiences, we might find exceptions, points of tension, inspiration and more questions as we go.
>
> *J* : How did I get here? It was actually a species of bird. I was in Trish's birding class at the time and I had heard from another student about a protest at Geprags Park in Hinesburg, which was the site of the last 2,000 feet of the ANGP. Geprags is Hinesburg's only public park, as well as nesting habitat for rare golden-winged warbler songbirds, which are in serious decline. The more than 200-person protest co-organized by 350Vermont and Rising Tide Vermont involved occupying the ANGP construction site. Sixteen of us got down in the trench and physically blocked the pipeline. "Do it for the warblers" was what I told myself as the police chief shouted, "Anyone in the pit is going to be arrested!" We just stayed there and kept singing.
>
> I had seen my dad that morning and he handed me a dime and said this is for your phone call home. I have supportive parents, yet I was terrified. I didn't know what I was doing, but I knew that I had to be there. It was the first time in years that I felt like I did something meaningful.
>
> When I later took Trish's "Environmental Activism and Policy" course we had an assignment to give a 3-minute statement before a commission. At the time, the Public Utility Commission[8] pipeline hearings were contentious, with sing-in actions leading to counter-strategies of locking people out. I half-jokingly asked Trish and Rachel, "what if I bought shares in Vermont Gas and made my statement at their shareholders meeting?" I read their faces as, "you have to do this!"
>
> So, I bought shares in the parent company, Energir and began my research using Trish's guidelines such as identifying your target audience. I rode the 4:00 a.m.

---

[7] We brought Brian into this project later in the summer to provide additional details on select topics.

[8] Known then as the Public Service Board.

Greyhound up to Montreal alone to the shareholder meeting in a place that had fancy tiles and art that was probably worth more than everything I own. I was performing what I think Ingrid has referred to in her class and research as "corporate drag". I clearly didn't belong. A woman approached me in the lobby saying, "Hi, I'm Sophie, is this your first time at the meeting?" I said "Yeah" and she said, "well enjoy, have some coffee, have some pastries" and I said "okay," thinking she was nice. Then she came back about 15 minutes later and put both of her hands on mine and she asked "are you good?" I said, "yes," and shortly after that, everyone headed into the meeting.

Then I made three pretty horrifying realizations. First, Sophie turned out to be the CEO of the company and was emceeing the meeting. Second, I thought I had chosen a good seat in the second row with a view, but then every person sitting next to me stood up one-by-one to applause because they were candidates for director positions. Finally, the entirety of the meeting was conducted in French, which I speak very poorly. But I understood the agenda included a question and answer period. I waited and then stood behind the microphone facing the CEO to ask "what would it take to stop this pipeline?" because at that point in 2017, the ANGP was a month away from completion. I don't know if I have ever been that terrified in my life, but I told myself again, "do it for the warblers." For me, the warblers came to encapsulate everything that I was doing this for, the indigenous communities that are impacted by the fracking this pipeline requires, for the people who were trying to protect that park, for everyone that I met along the way. Sophie gave this very polite and greenwashed response about how Energir cares so much about their customers and everything that they do in Vermont, and that I was so brave for coming to Montreal. At the end of the meeting I started handing out flyers that Rachel and I had made with information about the pipeline's inflated cost and environmental problems. The shareholders seemed largely uninterested, but a business reporter with the Canadian Press approached and interviewed me. He wrote a piece that began with the environmental problems with the pipeline in Vermont [1]. I realized that there was something to this "corporate espionage" thing. I continued working with Protect Geprags Park on the research that led to the current investigation of the pipeline. I also continued to speak out at Energir shareholders' meetings. All of this eventually led to my job here at 350Vermont.

T : What was it about that little bird? It was a golden-winged warbler… and we didn't really talk about it in class that much. I don't remember you as a bird person.

J : No, we didn't talk about it much in class. I was a zoology nerd and wanted to pursue that. But then I learned about climate change and habitat loss and extraction processes, which threaten all of that. The warbler became emblematic of this pipeline fight, of Protect Geprags Park, of this grassroots group. I have always been drawn to animals and I was in a rough spot in my life at that point. It helped me to have something outside of myself to work toward, to protect. I think it could have been any species to be honest. It could have been some rare worm.

R : The golden-winged warbler is so emblematic because it really *should* be listed as an endangered species. A working group within Fish and Wildlife has argued that nationally, pipeline utility corridors are great habitat for golden-winged warblers and thus we should never block pipeline corridors for the sake of an endangered bird. They are kind of the clean green cheap gas bird.

I : In the Pacific Northwest, the spotted owl has become an enduring negative and positive political symbol for many people. With Protect Geprags Park, are there similarities in terms of strategy by making the golden-winged warbler so prominent? Are there place-based differences we should consider?

M : Well that's a totally different environment, right? The spotted owl issue is a lot about worker's rights. We don't have that issue with this pipeline because they imported workers from Wisconsin! In Vermont, many people are pro-environment, for the most part. There are lots of pro-gas people here too but putting the golden-winged warbler on all the lawn signs is a good strategy here.

R : I think there is an element even here in Vermont that saw that as being really silly.

M : Oh sure, yes. But it doesn't escalate to what you see out west with the spotted owl, where it's the logging industry pitted against the environmentalists. In this pipeline corridor, the workers are farmers, predominantly.

T : I have another question for Julie. I had never seen a student go so deep on an assignment in terms of follow-through and the number of sources, and it was because of the link with you, Rachel. It was because of the critical link with the local organization.

J : Well yes, that's what you had provided for me, Trish, someone who was already doing this work within the community. Rachel gave me good starting points such as the company name and places I could look for information, including circular documents handed out at shareholder meetings. A lot of the research had been done. I just had to read it.

I : Were you hesitant at all Rachel, to work with Julie? Because our students run the gamut in terms of level of skills and interest in these topics.

R : Julie just came and did. She didn't need hand-holding.

T : But why didn't she? I'm interested in replicating this because I don't want to send unprepared students to local organizations. This is why I train them and take them myself to the schools where we work with kids. Because otherwise it doesn't work. I want more Julies in my classroom, but this was unusual.

J : Trish, you and I talked about this after sending students to the group trying to stop the F-35 jets from being based in South Burlington. It didn't work because the students were doing it for a grade, not because they wanted to engage with the issue. For me, part of it was that I just really like getting in the weeds on this kind of stuff. But there has to be some kind of personal connection to a project. I wouldn't want to make "more Julies." I'm not interested in a hero narrative–I think the key is to provide spaces that allow students do something tangible as part of a community.

I : And you are not from Vermont? You are from out of state?

J : No. I'm originally from near Chicago, but I've lived here for 15 years.

M : I think that's a really important distinction. Unlike many UVM students, Julie has lived here in this place. This is something I have a lot of feelings about, as someone who did a lot of jumping around the country for higher education. I believe that this is truly a part of the climate problem, and I call it modern colonization. [When we send our kids away to college] we are separating young people from their communities where they have the most agency and power, shipping them around and they don't have a connection to place. Julie is kind of an exception to the norm. We've worked with a lot of students through our internship program, through service learning projects. We did a service learning project with Brian's class a year ago. Out of a group of six people there were maybe two that really had any kind of heart in it at all. It's a drain on our staff to support those students. [350Vermont's staff] ended up making a ground rule that we are only going to do one service learning class a year even though we get asked a lot. But even the students who have heart are often off to the next thing. Because I really believe this work [of organizing] is about relationships, it's hard to see how investing in students—from my perspective of leading an activist movement-building organization—is worth the effort if they are not going to be part of this community for the long haul.

R : I totally relate to that. The one thing I would say though, is if there is a way to design a little piece that people can come in to do, that works well. It is often hard to come up with that kind of thing.

J : Right. And in terms of what Protect Geprags is doing right now, I get asked, "how do people plug into that work?" Unless you want to sit there and read seventeen thousand documents, there's not an easy way in. The pipeline is complete. There is nothing to go get in the way of. There's not necessarily something to organize around. So, we have to create those opportunities like we did with the recent Next Steps climate walk where we followed the pipeline route to Geprags Park and highlighted what happened there [4].

T : I want to respond to your point about modern colonization because I've been thinking about this since I started working in academia. When I was getting a PhD in Madison, Wisconsin, I was deeply embedded in my community over eight years as an activist. But then I was forced to leave to get a 'real job' and I was lucky to get hired here at UVM. And I haven't been here four years and I still don't have the connections yet to do really meaningful work. I'm just building these relationships. I started a project in two of my local schools in my neighborhood. These two after-school nature clubs are small and they work. But they are based on short-term relationships. I think it is so important to question how we are part of that modern colonization; how can we disrupt it and not reproduce it because we are reproducing systems of oppression. I'm thinking of Paulo Freire [6], who we study but we don't practice, right? If it's not worth your effort as organizations to help educate our students, which I totally understand, what about the students who are from Vermont and who might be here for the long haul? For example, I am interested in getting more kids from our local schools into UVM. This is a justice issue because we are a "public school" [only

26% of UVM undergraduates were from Vermont in the fall of 2019, according to UVM].

M : I also want to counter my critique, too—we still do the work of engaging with students because I have a lot of faith that, like you were saying Rachel, there is a seed you are planting. Can we give them a little something, a little taste of what activism is, what collective liberation is? Having students as a part of 350Vermont's five-day climate walk sparked an "oh, THIS could be community," moment for some. I have to have faith that our investment in students has a ripple effect. We may not really know how that changes the world, but I have faith that it does.

J : Right. The action that 350Vermont organized at Geprags Park radicalized me to a degree…not that that's work that you ever finish.

R : Maeve said that fundamentally this work is about relationships and I think that's something that has to be explored too because I really believe that that's the case as well. Bringing a bunch of strangers together and saying "hey, we are going to go do…" doesn't work very well. I am going to go back and touch gently on what Julie said about having a hard time. People get into a hard time and they are looking for some meaning, something that feels real and purposeful and that's a place where people can be mobilized. You give them something that's going to be really helpful to them and to the movement. Because it's often the people who are struggling with that feeling of meaninglessness or whatever it might be. I don't know, but I've encountered this in other places with people who are just struggling…I think of activism in a lot of ways as being psychotherapy for people.

I : I have seen many students struggling with the lack of connection to place and relationships and this sense of impending doom. Some students find a place where they develop those relationships. But I also see another kind of more insidious direction, which is more about consuming an activist moment as a tourist-like experience. There was a bit of that going on at Standing Rock, with a few folks posing for photos in an 'I'm here to be seen being here' context. I also worry about the generalizations about college students floating out there. There are many students who treat college as a consumer experience and there are those who are having trouble finding their next meal, so this can add to the challenge.

T : You both talked a lot about relationships. But how do you develop it in a semester? And then students move on to the next project. Because that's the system we have put students in and the system we work in. Do you have more thoughts on how to disrupt this? I'm always looking longer term. I have students who repeat the birding class so they can continue working with the same kid. Because our students understand that it *is* about the relationships. They were the ones who came to me and said, "I want to repeat your class." And initially I thought, "What? Repeat the class?"

M : Can they get credit for that?

T : Yes, but this idea came from the students. It didn't come from me. I didn't think I should figure out how to make these relationships indefinite because that's not the way we [academics] think or the way the academic calendar is set up. And it's questioned. But it came from the students. I am looking for ways to change the structure to suit my neighborhood.

J : ...Many of these kids are coming from underserved populations. There's a lot of refugee and lower income families in the area, so this is also a kind of after school program for one day a week. These kids develop a really deep connection with this place and often with their mentor and, well it's a mixed bag...

T : Ha-ha! Your kid was a pain!

J : Yeah, but it was meaningful.

R : You shouldn't sugar coat it.

J : I shouldn't sugar coat it. It was tough. We had what I called a "working relationship," but we had our moments. The class holds this idea of environmental justice through this place and I could watch it before my eyes... kids developing a relationship with place. There were these stinging nettles everywhere and we told the kids that jewelweed was a natural remedy for nettles. Every Wednesday we'd run at breakneck speed through the woods, but the kids would slow down and say, "no wait, stop we can't step on the jewelweed." They grew to appreciate this plant because they realized that it was there to help them if they got hurt. The class continues for kids going to Hunt Middle School and then I think Trish's dream is to have it eventually go to the high school and then to college...with a scholarship program so that these kids that are, again, generally underserved, have this opportunity that other students at UVM have.

T : And it's in their neighborhood, the place is. The question is about place and agency. Because these kids can go and defend that place if it is threatened by development. That's what happened in Madison with the outdoor classroom in a park where I first started this mentoring program. The Parks Department was cutting down the trees in our outdoor classroom and the kids started going before the Parks Commission crying and screaming at city officials not to cut down "their" tree. So, the kids became agents [of change], like you said, in their own place; it was their agency and power.

J : Right. Every Wednesday we would go with the kids. But then two other days per week we were at UVM talking about systems of oppression and sense of place...and birds of course, some of the same topics I learned about in Ingrid's classes, like decolonization and extractivism and the kinds of things that you can teach to a small class but are hard to get into the mainstream that are critical to doing this work in a meaningful way that doesn't become just 'I'm here to be seen.' All of those classes in conjunction with Bindu Panikkar's class where I learned a lot about what privilege and race meant in the context of environmental justice provoked some resistance at first. Like thinking, 'how could someone live somewhere more polluted just because they have black or brown skin? I don't get that.' But then I learned what it meant for myself to be white or otherwise privileged to be participating in neocolonialism. It's hard stuff to learn but it is some of the most critical knowledge to have in order to do this work.

M : Trish, I loved that you shared your story from Madison, and it reminds me of a part of the film [2] based on Naomi Klein's [12] book, *This Changes Everything*. I believe it was on activism in Greece around a gold mine. And I just remember very clearly that a woman who was interviewed basically said, she had such a strong connection to place that she was willing to put her life at risk to stop the mine. That's what it's going to take for us to survive the climate crisis and fossil fuel extractivism. I feel like so many people in this country especially…not entirely everyone, but people of my class or my bubble…this kind of elite academic modern colonizer…. There are very few of my peers who I know would say, this place is so special to me that I would actually lay my body down. Whereas I think that's true for Geprags Park and this pipeline. There are people who feel so strongly about place here in Vermont and that were and continue to be part of that campaign.

R : A lot of it is about place. For many land owners it's about place. I am particularly in love with the place where I live, which I am very fortunate to have fallen into, and I imagine sometimes what I would feel like if somebody had come in here and bulldozed it…I would lay down and die for this place. I mean that's what we've seen with indigenous peoples in many places fighting to the death over deep connections to place.

J : And that did happen in Alberta, where our fracked gas is coming from Lubicon Cree First Nations land [8]. It was a similar situation to Standing Rock where they set up barricades and blocked construction, that's where our fracked gas is actually coming from…2,000 miles away. There is this assumption that, "Oh, Vermont is so green and so liberal so everything is fine." But there are these two competing things. Fracking is banned here. There's nothing to frack here, but "it's okay to import this stuff."

R : We are part of the fracking system

3.2.2 Practicing Pipeline Pedagogies Amidst Higher Education Institution (HEI) Politics

T : Maeve, you said in the beginning that students will have the most agency and power where they are from…In the second class that Julie took, where students write a commission statement and testify, they can turn it into an op ed to publish. Now I have many students who are going back to their hometowns to do this. I have an impact map of states across the country where they are publishing an op-ed to stop Styrofoam in upstate New York or pushing for a green roof on their high school in Massachusetts. They are trying to have an impact at home but they are learning those skills here. And they are learning them from people like you. So, the seeds you are planting *are* bearing fruit across the country…This is the challenge: how do we not burden organizations? How do we practice reciprocity and accountability as Laura Pulido [17] says?

J : Well, there is a degree of reciprocity because these classroom and activist connections give us a platform to get our movement and our stories out. When you invite me to come to your class and speak it's a chance for us to inform people about what's happening here.

R : It's also different because you go into a class and you do a talk and then you leave and the investment is not as much as supervising a student to do a project. But what I have experienced too, is that a lot of folks, not just students but others too, are wanting to do something but they just don't know what to do or how to do it…sometimes even the simplest of things like learn to write an op ed.

T : Basic skills. And the other problem I have with teaching in a way that builds relationships—which is absolutely baseline for me—is that it goes against corporate university trends that measure everything in numbers. A college student and a kid the first week holding hands when they are walking back from our outdoor classroom…How do I measure that? I know from a student's paper that their mind is blown. They can't believe the kid trusts them enough to grab their hand and hug them and all the other little magical things that happen. But then universities pressure you to increase your class size…What does that do to relationships? Relationships are not valued… How do we defend relationships at every level?

R : You have to figure out how to make money out of building relationships Trish! *[all laugh]*…

M : A love pipeline… *[all laughing]*

I : That's another thing that I think some of the other participants in this volume are struggling with in terms of these dual pressures. It is a privilege to be teaching at a university, although that position is not the same for everybody. I have colleagues who teach eight classes a year and do not receive a livable wage for our county, while others are much more comfortable. There is a whole range of what that labor looks like in an institution, and then if you want to do something like create radical pipeline pedagogies it's often done as extra work, unseen or penalized for not channeling your energy more "productively." To what extent in your own work have you seen universities—big and small—at least in Vermont be helpful or a hindrance to your goals?

R : I think Maeve might actually be better able to talk about the ANGP and the university's role in that because we've had a very specific sort of niche of what we've been doing at Protect Geprags. But in my other work with Biofuelwatch, I really get angry at academics who get all these grants to engineer crops, trees and microbes to make bioenergy, which is absolute nonsense and not only doesn't work, but also creates more deforestation and industrial agriculture and results in competition with food production. These academics do it because it just brings in big grants. There's like three academic researchers in the entire country who do occasionally say "Hey, this is a stupid hoax. It's never going to work and you are wasting your money and time." They are all feeding out of the trough from USDA and DOE grants. So, I have a bad attitude about that. I think some of these academics do students a huge disservice by advocating for false solutions like biofuels. Students come out of their courses with a lot of misinformation and false hope about technofixes that do more harm than good. But it takes a lot of time and research and critical thinking to really sort out the 'wheat from the chaff' when evaluating these technologies and different proposed 'solutions'.

M : Trish, you brought a dozen students to the hearing that we had just a month ago, and that's really helpful. We are getting youth voices in front of legislators and we are packing the room. Even though it's a one-time thing, I think something like that is helpful. I don't have any examples of hindrance necessarily for the pipeline project other than there could just be so much more coming from the university. UVM is the 'green' university, and yet here's the biggest buildout of fossil fuel infrastructure this state has seen in years and not a peep. There were some students and professors of course, who were engaged, but not a huge institutional campaign. Maybe we could have pursued that.

R : Although, Rising Tide came out of the university. Some were from UVM, Middlebury…

I : It's eye-opening comparing Sterling College with Green Mountain College with Middlebury with UVM, etc.

J : There are some movements that are starting now…colleges like Middlebury are latching onto the Sunrise Movement, Green New Deal types of things and we have some students coming into work with us on that front but, most of what I've seen in terms of universities "going green" has been a lot of greenwashing. Like, "Well, all of our buildings are LEED certified," it's more about marketing than about actually taking steps toward achieving something bigger. Even within Burlington there is a lot of greenwashing; they say, "we get a lot of our energy from Hydro Quebec," which is itself incredibly problematic and extractive and damaging to the indigenous people who live in Quebec. It's very surface level.

M : Well, like Middlebury College wanted the gas! [14]

I : Yes! I was trying to get us there…

M : I would make a real distinction from what you are saying Julie around the students that come out of universities and often are engaged in activism where the target is the institution, like the Sunrise folks came out of the divestment movement, which was many years long and finally has had this big win, with Middlebury, so that's a real change, definitely…versus institutions like Sterling College and their president, Matthew Derr being an ally and divesting. I don't know if they even had anything to divest, but saying "we are an institution and we are taking this action as an institution," that at least shows some solidarity.

J : I didn't make an effort while I was at UVM to get involved with student groups. I just latched onto Protect Geprags and did that work because it was more meaningful and there was less red tape. I didn't have to get things sanctioned by the Student Government Association…

M : It has been a struggle for us too. With the pipeline campaign, thus far we have had internships and have worked with allied professors. But with the divestment movement, we had a field organizer who was doing campus-based organizing and working with lots of different schools. But again, that was tricky. There wasn't a deep relationship there between the field organizer and the students and there was high turnover in that field organizer position. Also, there were some really strong student leaders that were driving that campaign at UVM and then they graduated and then there weren't strong leaders, and the divestment campaign suffered. What can we do as an outside organization to try to sustain a campaign like that? I don't feel like we've figured out the best structure. I know VPIRG[9] has had a student chapter, but it seems oriented to recruiting students for their summer canvas.

R : You know, I come out of academics too, and one of the things I often come up against is that I'm very research oriented. Biofuelwatch as an organization basically does research and then we try to do the groundwork to support campaign work. There have been tensions within the pipeline movement at different times about strategies. People have argued that nothing's going to work other than chaining our bodies to equipment or NVDA. Protect Geprags Park has been very research oriented. But it's hard to be inclusive when you're doing a lot of research because many people don't have experience with that or don't want to do it and it's painstaking and time consuming. But it's also a way to get people really engaged…especially university students, because once you start to learn about it and you dive into it in that more detailed way, you start to own the knowledge and understanding. Really, doing research can be a pathway "in" to an issue…Activists in general need to embrace it more I think. So, that they are not just up there saying, "Fix the climate, but really having a much deeper understanding of what that means.

I : I really resonate with that. I love reading and listening to the voices of the civil rights movement. All the people and kinds of labor that it takes to pull off a march…to provide sustenance, first aid, creative signs, etc. There are so many roles…for introverts, for people with chronic illnesses. I've thought a lot about that in terms of what activist practice looks like from different embodied perspectives [see 11, 19]. What if you can't be at the march and put your body on the line? There is some romanticizing of activist superheroes here in Vermont and in other places [see 15].

R : That's really helpful for relationship-building too, because you find people who can do the same kinds of tasks and those are the people you can work with.

T : …The challenge I see is, how do you break a research project up into pieces, to give a student a piece that they can actually focus on and do? Because the nature of research is you go, you dive in and you start to go, go, go, go. How do you supervise that?

J : All of the pieces are important. Phase II of the pipeline was largely canceled due to the public outcry and lack of funding. The current investigation of the pipeline is all based on research and photographic evidence… and that's the closest we've ever gotten to actually shutting this down. It is immense amounts of research and legal challenges done by folks from Protect Geprags.

T : And that's the case abroad. The 2011 suit that they won in Ecuador. It was because of evidence and research. Could organizations here put out a list of research needs? Say: "here's ten topics we need undergrads to research. We need to know about this pesticide or this pipeline or we need a bio on this CEO." [Give us] ten research needs a semester and we give them to students. The challenge

---

[9]https://www.vpirg.org/.

will be the quality of work. But that's our job to train them to do a decent job, grade them and tell them to revise it.

R : Well, for example, people who worked with Protect Geprags have done a lot of public records requests. Just learning how to write a public records request, it's a skill! For a newcomer it's like, "oh, I can get access to this kind of information?!"

J : And the stuff that you find is fascinating...

R : Yes, definitely we could come up with lots of research questions. Personally, I came out of a decade of full-time motherhood. I had been in academics, and then was trying to figure out what to do. I went to VPIRG and volunteered for them. I was working on lead and toxics in toys...just started researching and the more I learned, the more it really mobilized me. It was empowering to be on a path of starting to know something about it. That itself can be really motivating. But you need also to have a plan to do something with what you are learning.

M : Yeah. I feel like we could probably come up with research topics pretty easily. It's been great to have Julie on staff, she's someone who can bring some of that research depth because of her work with Rachel and Protect Geprags Park. I think a field organizer who has a research strength brings a certain type of asset, and can really dig in when I say, "I want to know everything about this person." We might want to know all about the president of VGS, for power mapping reasons, to know their connections and pillars of support. When we think about doing brand busting with VGS, that would be really helpful.

J : One of the disconnects for me was that I had a strong research background, but then doing something with what I discover is difficult. I'm now in this position where I have to plan a campaign, and I'm still learning. I wonder if there's space in higher education to teach something about how to build a campaign or if that's too radical?

I : I think Brian has done some of that in his classes.

T : But the nature of what is a campaign is changing so fast with social media. So, I have case studies and I was involved in campaigns and I know something about it, but that was 10 or 20 years ago! What is a campaign now? What does it look like? You [Julie] would know more about that.

I : The strategies also have to adapt...corporations get used to this tactic of people becoming shareholders and crashing their meetings and then they come up with new, well-resourced counter-strategies. Activists constantly having to reinvent the wheel can be exhausting.

R : I think one of the things that plagues activism is in-fighting. People end up bickering with each other over how to do it, what to do. Who is purer in their philosophy or whatever and stepping on each other's feet. It is our downfall.

T : ...Fractions, splits, conflicts, tension and out of that comes new things and new ways of doing things.

R : Or failure.

T : Or failure. That's true. Divide and rule...

R : But I think that kind of thing has to be part of what's taught in campaign work. It's messy, it's hard to do and people disagree and have issues and even fight amongst themselves. There's a lot of divide and conquer that is used against us and we're often unaware of it. You know, it's not really just about writing op eds and campaign signs, it does go back to that relationship thing. That also needs to be taught! In many ways we have in recent history been "unlearning" how to be in personal relationships with one another.

T : Something that I find fascinating and horrible about the Vermont pipeline story…It's like the police brutality issues right now in Burlington [see 18]. It's the maple-syrup-coated beautiful Vermont story people have in their heads. And then there is this other story of information hidden from the public, of processes that seem democratic and they're not, like the Public Utilities Commission shutting people out of public meetings. This is the real Vermont, the other Vermont and maybe that's why there's been some resistance to covering the story or teaching about the pipeline…all of the university's business interests.

R : Well and gas, people have this story about gas as "the bridge fuel" that is cleaner and greener even though it is certainly NOT.

J : And "renewable natural gas," which is the next bridge fuel myth and greenwashing scheme to keep us hooked on fracked gas [16]. It is the physicality of it. The pipeline is this out of sight, out of mind thing for many people. With Geprags, the Supreme Court of Vermont said that because the pipeline is under the surface of the park, it doesn't materially change the use of the park as a park. So, it's fine. It's compatible with the public park because you can't see it. That was basically their argument. There's constantly this question of "now what?" with every setback, which goes back to that idea that these things are always evolving

3.3 Re-Connecting with Brian After Our Multi-logue

Our colleague, Brian Tokar (B), was not initially part of the recorded multi-logue, but we felt his absence and agreed that we wanted his perspective on working with students and climate activists regarding the ANGP. We invited Brian to address specific questions that arose, not to comment on the dialogue as a whole. His comments add details on the nature of student projects and research in his classes, the origins of Rising Tide and other activist campuses in VT, and the marginalization and support he has experienced from UVM.

3.3.1 The Nature of Student Projects and Research

B : My experience confirms that students need some hand-holding along the way. In my classes I start with a general brainstorm of project ideas in the second or third week, followed by a second go-round a couple of weeks later, where students begin to commit to specific projects and form groups. I make time for regular check-ins through mid-semester, when each group needs to submit a proposal. They know it can be informal and won't be graded, but that I will take the opportunity to offer specific comments, suggestions and community contacts. There's also some in-class time devoted to group meetings to supplement their out-of-class meetings and offer further assistance.

Less activist-oriented students typically do surveys on various topics, film showings on campus, and other similar activities. Students who are interested in activism seem to especially appreciate the chance to see what it's all about in a semi-structured setting. For many, it's definitely a "taste of what activism is," as Maeve mentioned, which can help feed a desire for more. Student projects related to the pipeline were wide-ranging, from fundraising events to art shows, participating in event organizing to interviewing participants at major rallies. I remember one student who made an exceptionally well-edited video of the gathering at Geprags Park where Julie first got involved. I asked her where she learned to become such a skilled video editor and she replied that she had made skateboarding videos all through high school! Clearly some are more engaged than others and there are always a few who seem to be going along for the ride (and the credits). But I've seen many past students at subsequent events over the years, saying their class project got them interested in doing more. The group exercises and support through the semester can also help with their inner group dynamics. I've had several students say at the end that they'd always dreaded group projects in the past, but that this experience was different.

Regarding student research, the last couple of springs I've also offered a Service Learning class, with more structured, semester-long projects in service to specific local organizations. Two years in a row, the most research-intensive projects were the most successful. First, investigating the potential effects of a proposed Vermont carbon tax on various household demographics, and second, updating a statewide handbook of town-initiated energy projects. I'm not sure why these were the strongest projects. It may be that the more career-oriented students in this upper level class are drawn to those projects and bring a higher level of focus. Projects focused on campus organizing have been somewhat disappointing in that class so far, as Maeve mentioned.

3.3.2 *On the Origins of Rising Tide, and Other Activist Campuses in VT*

B : The original five founders of Rising Tide Vermont were all UVM grads including three Environmental Studies students, and one Middlebury College student who became a core group member a year or two later. Several students from the Environmental Program became active members over the course of the pipeline campaign, and others got involved in the campaign through organizations like VPIRG. Middlebury also has a long history of environmental activism on campus—my recollections go back to the campaigns against the expansion of Hydro-Quebec's mega-dams in the early 1990s—and of course it continued through the development of 350.org with Bill McKibben, and more recently an active chapter of the Sunrise Movement,[10] focused on promoting the Green New Deal. Middlebury has had informal meetings of environmentally-minded students every Sunday night for at least a couple of decades, and those gatherings have formed the nucleus of numerous important projects. Johnson State—now part of Northern Vermont University—also used to be an active campus, and I'm not sure what's changed there. Sterling College, which is very focused on experiential education with concentrations in agriculture and forestry, has become increasingly active and was also a big presence at several of the pipeline protests. The last time I offered my climate advocacy class at UVM, two groups of students developed projects around the idea of UVM "greenwashing." One group interviewed several students, professors and an administrator or two, as well as a local architect who had taught at UVM but became very disillusioned with UVM's building practices (e.g. their flaunting of USGBC LEED certification for buildings that only met the letter of the certification rules). The other group tried to influence the university's selection process for a new president, and found their efforts were stymied at nearly every step (not surprisingly). They eventually organized a petition drive asking the search committee to prioritize candidates with a demonstrable environmental commitment. Instead, the UVM Board of Trustees hired an engineering dean from Purdue whose most flaunted specialty was in managing corporate partnerships.

Regarding VPIRG, it's true that their main focus on campus has been recruiting students for their summer canvass. It started out as a student organization in the 1970s, which was a long time ago in the academic world. There have been a couple of attempts to organize students on behalf of VPIRG and bring UVM students into a statewide climate advocacy network initiated mainly by high school students, but they've not been very successful so far.

3.3.2.1 Regarding Marginalization and Support from the University

B : I've had tremendous personal support and encouragement from the faculty and leadership of the Environmental Program since I came to UVM in the mid-2000s, but that has only minimally translated into real institutional support. As a part-time faculty member I am not eligible for any of the grant support offered through the Gund Institute for Environment.[11] My course-load is fairly consistent, but I've seen many other like-minded part-timers dropped from the faculty, especially since UVM shifted toward an Incentive-Based Budgeting model, which forced various colleges and departments to compete for "student credit hours." Instead of my classes making money for the Environmental Program, as they did when they were under the auspices of the Continuing Education program, they apparently

---

[10] https://www.sunrisemovement.org/.
[11] https://www.uvm.edu/gund.

have to scramble now to find the funds to pay me to teach. I do all my research and writing on my own time, with no support at all from UVM.

3.4 Concluding Remarks—Moving Forward

Through the course of our conversation about pipeline pedagogies, it is clear that there are many spaces and practices of pipeline resistance. From the more obvious public meeting spaces and sites of pipeline construction to shareholder meeting rooms, classrooms, neighborhood wooded areas where birds and kids meet and hours spent behind computer screens. Maeve and Rachel's frustrations with service learning classes that deplete the energies of local organizations and organizers are important. We hear their call for teaching that gives students the concrete research, thinking and writing skills that local organizations need—instead of expecting the organizations to do the teaching for the university. This is critical for helping students become not activists in the "activist superhero" mold or "consumptive-Instagram" activist mode discussed in our multi-logue, but citizens of our cities, rural spaces and beyond.

It's also important not to see pieces of fossil fuel infrastructure in isolation. The ANGP is just one small piece of infrastructure spanning across all of North America leaving points of destruction in its wake. The fossil fuel industry's influence extends far, but the resistance reaches even further. We could delineate a "frontline" here in Vermont along the pipeline's right-of-way, but resistance happens on the floors of churches where late-night, last-minute action planning happens, in the corporate headquarters of the fossil fuel company, the homes of activists hunched over mountains of jargon-saturated documents, hell-bent on figuring out how to save these places we love…and by extension the beings that live, have lived, and will live there. We embody these struggles and bring that sense of "place" with us.

References

1. Arsenault J (2017, March 21) Projet De Gazoduc Au Vermont: Gaz Métro Fait Consensus, Dit Sa Dirigeante. La Presse. Retrieved from https://www.lapresse.ca
2. Barnes J, Lewis A, Cuarón A (Producers), Lewis A (Director) (2015). This changes everything [Motion Picture]. Abramorama, Canada and the United States
3. Bosworth K (2019) The people know best: situating the counterexpertise of populist pipeline opposition movements. Ann Am Assoc Geogr 109(2):581–592. https://doi.org/10.1080/24694452.2018.1494538
4. Box O (2019, April 29) Vermonters marched 65 miles for climate justice. In These Times. Retrieved from http://inthesetimes.com/article/21854/vermonters-march-65-miles-climate-justice-ban-fossil-fuel-pipeline-350VT
5. da Costa LB, Icaza R, Talero AMO (2015) Knowledge about, knowledge with: dilemmas of researching lives, nature and gender otherwise. In: Harcourt W, Nelson IL (eds) Practicing feminist political ecologies: moving beyond the 'green economy'. Zed Books, London, pp 260–285

6. Freire P (1970) Pedagogy of the oppressed (translated by Myra Bergman Ramos). Herder and Herder, New York
7. Gallagher NL (1999) Breeding better Vermonters: The eugenics project in the Green Mountain State. University Press of New England, Hanover, NH
8. Goddard J (1991) Last stand of the Lubicon Cree. Douglas & McIntyre, Vancouver
9. Hall F (2015) Deconstructing systemic oppression through teaching community organizing: a students-teaching-students course. Undergraduate thesis, University of Vermont
10. Haraway D (2016) Staying with the trouble: making kin in the chthulucene. Duke University Press, Durham, NC
11. Hedva J (2016, January 19) Sick woman theory. *Mask Magazine*. Retrieved from http://www.maskmagazine.com/not-again/struggle/sick-woman-theory
12. Klein N (2014) This changes everything: capitalism vs. the climate. Simon & Schuster, New York
13. Landen X (2018, November 20) Group asks state to expand investigation of Vermont gas pipeline construction. VTDigger.org. Retrieved from https://www.VTDigger.org
14. Liebowitz R (2013, May 6) Middlebury college statement on natural gas pipeline. Middlebury College Newsroom. Retrieved from http://www.middlebury.edu/newsroom/node/450619
15. Nelson IL (2015) Feminist political ecology and the (un)making of 'heroes': encounters in Mozambique. In: Harcourt W, Nelson IL (eds) Practicing feminist political ecologies: moving beyond the 'green economy'. Zed Books, London, pp 131–156
16. Powell T (2019, February 12) Calling natural gas a 'bridge fuel' is alarmingly deceptive. Sightline Institute. Retrieved from https://www.sightline.org/2019/02/12/calling-natural-gas-a-bridge-fuel-is-alarmingly-deceptive/
17. Pulido L (2008) FAQs: Frequently (un)asked questions about being a scholar activist. In: Hale CR (ed) Engaging contradictions: theory, politics, and methods of activist scholarship. University of California Press, Berkeley and Los Angeles, pp 341–366
18. Quigley A (2019, May 3) Burlington officers named in lawsuit shown pushing, tackling in body cam footage. VTDigger.org. Retrieved from https://vtdigger.org/2019/05/03/burlington-officers-named-lawsuit-shown-pushing-tackling-body-cam-footage/
19. Ray SJ, Sibara J (eds) (2017) Disability studies and the environmental humanities: toward an eco-crip theory. University of Nebraska Press, Lincoln
20. State of Vermont Public Utility Commission (2020) 17-3550-INV Regulatory case details (All other documents filed by parties resource tab). Retrieved from https://epuc.vermont.gov/?q=node/64/111907/FV-ALLOTDOX-PTL
21. Wiseman FM (2001) The voice of the dawn: an autohistory of the Abenaki Nation. University Press of New England, Hanover

Chapter 4
We Are Teachers and Learners Together: Cross-Disciplinary Lessons from the Pilgrim Pipelines Dispute

Lisa Jordan

Abstract Pipeline disputes present an opportunity to engage students in real-world contemporary environmental problems that relate directly to coursework in environmental studies and sciences (ESS). Using my own involvement with the Coalition against the Pilgrim Pipelines (CAPP) as an example, I demonstrate how community-based learning relates to course material in public health, Geographic Information Systems (GIS), and introductory environment and society courses. By threading together my advocacy work across courses, I express the ways that I came to encounter and value local knowledge and relationships. I also highlight the role of institutional arrangements in the support and promotion of community-based learning (CBL) pedagogy.

Keywords Pilgrim Pipelines · Geographic information systems · Community-Based learning · Local knowledge · Bakken shale oil

The purpose of this chapter is to describe the use of pipeline disputes as a teaching instrument for community-based coursework in environmental studies, geographic information sciences, and public health. By adopting CBL pedagogical approaches, I learned to appreciate local activists, local journalists, local political actors, and their knowledge and experience, as important partners in both teaching and learning. I feel the CBL approach required some unlearning of biases connected to my own education: valuing national and international-level surveillance of trends, dismissive feelings toward local journalism, privileging large-scale transformation and data-driven recommendations, and promotion of hierarchical sensibilities, which implied that academic sources are inherently preferential.

Drawing from my own experiences as an activist, community member, and teacher, I describe how I became involved in a local pipeline dispute, the various actors that I met, the opportunities that community and service-learning programs

L. Jordan (✉)
Department of Biology, Drew University, Madison, NJ, USA
e-mail: ljordan@drew.edu

provided in connecting activism to teaching, and the lessons that I learned from this process. I begin by introducing the different conflicts surrounding the proposal to construct a bi-directional, 170-mile pipeline that would carry Bakken Shale oil from Albany, New York to a refinery in Linden, New Jersey, and the stories of the activists that I met who worked to draw attention and concern to the project. I outline the various experimental teaching strategies that I adopted in bringing information about the dispute to students, such as attending public meetings with students, conducting Skype meetings and in-class visits with activists, preparing community asset maps, and creating story maps of the potential disruption to natural and cultural systems. I conclude that experiences with advocacy and teaching surrounding pipeline disputes provide helpful examples that could be applied to other forms of environmental disputes, such as community involvement in Superfund sites, brownfields, urban and regional development projects, or concerns about drinking water quality.

In this chapter I connect my experiences to three applied scholarly fields: work in health communications [44], research in place-based community engagement [53], and advocacy through teaching in environmental justice [5, 13, 26, 33]. These connections reveal insight into how pipeline curricula create spaces that value local knowledge, how witnessing local political negotiations presents opportunities for meaningful individual civic engagement and participation, and how students can serve in their own interest while organizing within larger groups that can provide greater leverage toward desired outcomes. The chance to personally see both local disputes and the successful organizing of coalitions in support of native peoples, such as the Ramapough Lenape Nation, demonstrated the complexity familiar to many developmental or energy-related planning projects, but also the opportunities for resilience and creative forms of resistance, where the university is situated in a humble role: one among many actors.

4.1 Background

Beginning in 2014, news sources began to follow a proposal to build two pipelines between Albany, New York and Linden, New Jersey [38, 51]. The pipelines would transport up to 200,000 barrels per day of Bakkan crude oil from the Port of Albany to Linden for processing, and return a variety of products, including gasoline, diesel, heating oil, and jet fuel [51]. The proposal was likely a response to substantial increases in oil production in Canada and North Dakota: 20–25 percent of Bakken crude oil, a light crude oil, unlike heavier tar sands crude, travels by rail, to the east coast [32]. Railcars carrying Bakken oil from North Dakota and Canada arrive in Albany and are currently shipped to New Jersey by barge or rail [42]. The stated purpose of the project is to "provide the region with a more stable supply of essential refined petroleum products" (Pilgrim Pipeline Holdings LLC) [36]. By the fall of 2014, the Coalition against the Pilgrim Pipelines (CAPP), consisting of over 40 organizations, had organized a domain (StopPilgrimPipeline.com) and began education and outreach campaigns in New Jersey and New York towns along the proposed pipeline route [30].

I became aware of the pipeline proposal after attending a local meeting held at Madison High School in Madison, New Jersey. The meeting filled the auditorium, and speakers at the meeting included the mayor of Madison, Bob Conley, several other local leaders, an environmental lawyer, and a very dynamic and well-spoken member of Chatham Citizens against the Pilgrim Pipeline, Brendan Keating [25]. The energy, depth of argument and research, and the level of engagement by the community took me completely by surprise. I had never been involved in a public assembly like this. I did not expect that so many community members would gather to learn about the possible construction of an oil pipeline.

I learned that the proposed pipelines would be built through our town, less than two miles from the university. I felt that the proximity to the pipeline, the energy from the community, and the sophisticated approach that opponents were taking to communicate their concerns might make this example a useful case study for my community-based learning courses. I was also very interested in the development of both the project and opposition as a researcher and community member. In addition to teaching in Madison, I live in Madison. I was not only meeting activists and political leaders, but I was meeting new neighbors, and people who would become friends. From these personal connections, I felt the benefits of place-based teaching practices, as described in the CBL literature [53].

4.1.1 Connecting Academia and Activism

Before becoming a faculty member at Drew University, I was largely disengaged from politics. Sadly, my own high school civics and even college education led me to the conclusion, for many years, that voting was the only form of civic engagement. Despite years of learning and teaching in a variety of fields, even, embarrassingly, environmental justice, it was not until I encountered the service-learning curriculum that I finally came to understand what real community engagement meant. Before I adopted CBL strategies, I had believed my research was "applied," as opposed to theoretical, because the conclusions might make a reference to policy implications. Furthermore, I felt the practical nature of my research, through its intended use for others, was important. But, CBL helped transform my understanding of how communication between academics and policy-makers takes place.

I first encountered environmental justice literature through the work of Robert Bullard on environmental racism [3] and the *Toxics Waste and Race* report (Commission for Racial Justice [10], when I was in college. I held these examples in the highest esteem because they seemed to me to demonstrate the purpose of research: the provision of compelling information and arguments to create policy change for the common good. The work of these scholars and leaders clearly demonstrated that exposures to environmental toxics were not evenly distributed across the US population, but that poor and minority communities were disproportionately affected.

However, I did not fully understand the way that research informs policy until much more recently. The piece of the story that I missed, for over a decade, was that the *Toxics Waste and Race* report was taken and used by community activists across a variety of cities and states to make the case for environmental justice [29]. Instead of relying on the national analysis as a whole, WE ACT, in Harlem, picked apart the New York data to lobby for changes in NYC and the state. It was these acts of local activism, taking place across the country, that translated the academic report into public demands and the eventual passage of Executive Order 12898 on environmental justice. The important connection I had lacked was an understanding of the role of the public, and specifically environmental justice advocacy, in promoting education, outreach, and eventually policy changes related to environmental inequities.

After graduating with an PhD in geography from the University of Colorado, I taught at Florida State University as joint faculty in Geography and Public Health for five years. Then, I joined Drew University as faculty in Environmental Studies and Public Health, directing the Spatial Data Center. I would have never become an activist-teacher if it had not been for a civic engagement program on my campus [47]. Drew University belongs to Campus Compact, a large coalition of universities committed to civic education. The purpose of Campus Compact is to advance "the public purposes of colleges and universities by deepening their ability to improve community life and to educate students for civic and social responsibility" [8]. Student enrollment in the campus civic engagement program, through scholarships and enrollment in courses that support an undergraduate certificate program, led to my initial involvement.

For me, understanding service learning came slowly. I learned by inferring from the examples of other faculty. In our program, faculty either identified their own community-based learning partnerships, or built from relationships previously developed by the Center for Civic Engagement.

Civic engagement courses require 18–20 h of service per semester from students, so I needed to learn which partners would make sense for me and the learners in my courses. First, I collaborated through the Environmental Protection Agency Toxics Release Inventory (EPA TRI) University Challenge [49], from 2013 to 2015, which helped me to understand much more deeply the Emergency Planning and Community Right-to-Know Act (EPCRA), as well as the EPA Toxics Release Inventory [49]. The TRI University Challenge was a volunteer EPA program developed to connect EPA staff and college faculty, and I was eager to take advantage of it. My early GIS research in graduate school as a geographer explored inequities in toxic exposure in New Jersey using the TRI database [28].

The EPA Toxics Release Inventory (TRI) followed from the 1986 passage of the Emergency Planning and Community Right-to-Know Act (EPCRA), which requires industry to list the potentially hazardous chemicals presently stored, used, and released in industrial processes [50]. TRI reports are completed by industry, and require an inventory on over 600 toxic substances. By meeting with EPA administrators, I was able to better understand the history, experience of data collection by EPA staff, and the really important and valuable information collected annually from polluting industry.

My previous experience with the TRI database had been as records in a spreadsheet. From Nora Lopez, the Region 2 EPA TRI coordinator, I learned so much more about the data and its collection. She drove to the companies in her car, and updated the GPS data in the TRI with her handheld receiver. She began work for the EPA prior to collection of the TRI. She recalled early meetings informing industry leaders, mostly men, about upcoming data collection requirements. As she told it, they held out their keys to her at the meetings to take them: they felt the EPA was taking away their businesses. During George W. Bush's administration, an optional short form for the TRI was introduced, to reduce data collection. Nora told industry to go ahead, complete the short form, and the EPA will audit you. I remember thinking how lucky New Jersey was to have Nora, and I wondered if other regions would be so lucky.

Now retired, Nora Lopez, splits her time between Puerto Rico and New Jersey. She explained to me that community organizations, like the Ironbound Community Corporation (ICC), worked on behalf of the communities to help communicate environmental information. Her affection for this group made me curious to learn more, so I learned about the ICC, and I attended several community action group meetings for the Passaic River Superfund site where the ICC was present. The heavily contaminated lower eight miles of the Passaic River constitutes one of the largest Superfund sites in the US. Where the Passaic River meets Newark Bay, the legacy of discharges from the production of Agent Orange by the former Diamond Alkalai plant have poisoned river and bay waters, aquatic life, and sediment.

From Community Action Group (CAG) meetings, I learned about Riverkeepers, both local and worldwide, who are environmental advocates for rivers and waterbodies. I took the eco-cruise along Newark Bay to learn about the many and varied environmental challenges that our waterbodies face [22]. Work with the EPA also connected me with the Deep South Center for Environmental Justice, and the New Jersey Department of Environmental Protection (NJ DEP). I learned about the New Jersey Environmental Justice Alliance, and the incredible work of Cynthia Mellon at the Newark Environmental Commission and environmental justice leader Ana Baptista, director of the Environmental Policy and Sustainability Management Program at the New School, who helped introduce and study cumulative impact environmental justice ordinances [1, 23, 31]. I was able to easily connect their work and recent engagement experiences to environmental studies, public health and my courses on mapping (Geographic Information Systems). It was in the second year of collaborating with EPA TRI University Challenge that I learned about the Pilgrim Pipelines and began contemplating ways to integrate the pipeline dispute into service-learning coursework.

The EPA offered an essential starting point for CBL pedagogy, in my own experience. Though the EPA TRI University Challenge has been temporarily suspended, a new program through EPA Alumni connects EPA staff, acting and retired, to faculty interested in bringing staff expertise into the classroom [17]. I have found this program valuable. For example, a recent environmental studies course that I taught benefitted from EPA staff contributions to class discussion on the challenges that the EPA faces in monitoring Department of Defense sites on the National Priority List.

4.2 Pipeline Coursework

I was able to use the Pilgrim Pipelines dispute in three courses, which I highlight next. I first adopted the pipeline case study in my Medical Geography course in 2015, which was a public health elective and CBL class. I experimented with combining the pipeline dispute, health communications theory, and social media outreach as a part of my Medical Geography course in 2016. I used the pipeline dispute as a community mapping exercise for the course *Environment, Society and Sustainability*. Last, I used a pipeline impact assessment as a Story Map assignment for Geographic Information Systems.

4.2.1 Service Learning in Medical Geography

After one year of collaborating with the EPA, I continued the TRI University Challenge, but I also began accepting invitations to collaborate with local activists on a variety of issues: Global Advisors on Smokefree Policy on smokefree campuses, Food & Water Watch on banning single-use water bottles on campus, the Sierra Club on local issues, the local environmental commission, and Wind of the Spirit, an Immigrant Resource Center. I continued my more established work with USAID Famine Early Warning Systems Network (FEWS NET), and benefited from a better understanding of how food security analysis operates alongside advocacy by Action Against Hunger, Humanitarian Open Street Maps, GIS Corps, and among institutions designed to help coordinate activities, such as the United Nations Organization for the Coordination of Humanitarian Affairs (UN OCHA) and their work on Humanitarian Data Exchange (HDX). This expanding scope of involvement often happened accidentally, meeting alumni interested in engagement, following up with the work of former students, or returning favors for the many people who were helping to teach my courses.

In the spring of 2015, I was asked to offer my public health course on Medical Geography as a service-learning course. In this course, I decided to include key topics that I thought were important to the field and would cover a reasonable breadth of information. I chose to focus on three areas within medical geography: global health, environmental health, and health behaviors. For each section, I chose subtopics and readings, where students collaborated in groups to prepare synthesis reports that were effectively policy briefs for community partners. At the end of each section, students presented their work to community partners. For global health, I paired my students with FEWS NET staff, for Environmental Health we worked through the EPA TRI University Challenge, and for health behaviors we worked with a tobacco control non-profit (NJ GASP). Students enrolled in the course had to complete an additional 18–20 hours of service outside the classroom.

For the environmental health section, I assigned *Toms River: A Story of Science and Salvation*, by Dan Fagin. This reading aligned well with the TRI University

Challenge, exploring the toxic releases in Ocean County, New Jersey, and the various roles that the EPA, New Jersey DEP, and advocacy groups played over decades of contesting water pollution. Several students were from Ocean County, had family that had worked at the Ciba-Geigy plant, and helped elaborate on different aspects of the context of Fagin's Pulitzer award-winning environmental case study. In learning about environmental health, we enjoyed guest lectures from Nora Lopez, Region 2 EPA TRI coordinator, and Matt Smith, an organizer with Food & Water Watch. I met Matt at an early CAPP event, and he volunteered to come to Drew. In establishing team-based reports, we brainstormed on topics that interested the class, all of which needed to include EPA TRI data. One group decided to explore the relationship between industry documented in the TRI and the proposed Pilgrim Pipeline.

The student group's research, using GIS, CAPP pipeline maps, the EPA TRI databases, US Census OnTheMap, and data from the New Jersey state Office of Information, revealed that in New York and New Jersey, over 175 thousand people traveled to workplaces within one mile of the proposed pipeline, 0.5 million people lived within one mile of the pipeline, and 1.5 million people lived within three miles of the pipeline. They identified 42 TRI sites within one mile of the pipeline, sites which emitted a total of 2 billion pounds of toxic chemicals to the land, air, or water in 2013. Our work was presented to Nora Lopez, and we shared our synthesis reports with Carey Johnston, EPA staff at the national offices. Johnston helped us learn and understand the computer database for distributing Discharge Monitoring Reports (DMR), a tool for exploring water pollution. We also presented our work as a poster at the NJ DEP GIS Poster Contest in Trenton, New Jersey.

Several students from the course joined me in attending Coalition against the Pilgrim Pipeline organizing and informational meetings. These were held at a local public library, Library of the Chathams, and we met Chatham Citizens against the Pilgrim Pipeline activists, Sierra Club staff, Food & Water Watch organizers, local officials, and long-time community activists. These meetings often were Advocacy 101 sessions for concerned community members, and they provided specific information about the Pilgrim Pipeline, providing ideas and techniques for writing letters to the editor, or opinion pieces for local newspapers. They shared examples of letters we could write to legislators.

The types of concerns raised around the pipeline were wide ranging, from water pollution and explosions, to disruption of traffic and businesses, to exposure for sensitive populations, such as children at schools and residents of retirement homes located immediately next to the proposed pipeline route. I was impressed with the specific information that organizers knew and shared: the rules and regulations of the right-of-way of powerlines (the proposed pipelines would be buried under existing electrical infrastructure), and existing legal protections that could undermine pipeline development, such as proximity to endangered species, possible consequences for designated historical and cultural landmarks, threats of contamination of sole source aquifers, and protections for drinking water sources (e.g. New Jersey Highlands). Students working on the synthesis reports were able to follow advocate recommendations, learning from practice and legal scholarship on the effective areas to emphasize. Organizers and local officials against the pipeline also crafted and shared resolutions against the pipeline. These were encouraged as mechanisms to effectively advocate against pipeline construction.

The work of the coalition highlighted to me the need for an "it takes a village" mentality. As a single teacher, I cannot communicate the depth of information involved in advocacy that an entire team of citizens can. While I had encouraged team work in the classroom among students, it became clear to me that when students were able to work in teams in a community setting that the breadth of expertise and experience was wider, if only because of its greater diversity, in age, in backgrounds, and in the many, varied roles of all those involved. Being in a setting were the goals of the gathering are to make our community better, together, is substantively different from a gathering where learning objectives or grades are the benchmarks. I learned later in Alternatives to Violence Prevent (AVP) training workshops how to verbally express the process of successful community engagement, which I was witnessing through the work of CAPP: (1) we are teachers and learners *together*, (2) we come to our work as volunteers, (3) there is good in every person. Also, in the practice of communicating, it became necessary for me to learn to step forward, when needed, and step back, to let others share and guide community action.

4.2.2 Health Communications Exercises for Public Health

In the year after teaching my first community-based learning course on Medical Geography, I became more involved in CAPP. I found myself listening in on biweekly organizing calls. I met several faculty members, across a number of disciplines from other universities who were also involved in organizing. I canvassed at a mall with Sister Jeanne Goyette, a Dominican nun. I also attended late-night video conference training for volunteers interested in social media outreach, with about a half-dozen volunteers in attendance. Matt Smith, the Food & Water Watch coordinator for CAPP, described how we could each help maintain interest and content for the CAPP FACEBOOK page by posting once a week to share any new information related to the pipeline. I was intrigued that posting at particular times of day lead to more views: Matt suggested that we coordinate to post at 11 am and 3 pm.

By learning about social media outreach, I cultivated a greater interest in work on health communications. In my course, I required Renato Schiavo's *Health Communications* textbook. For Spring 2016, only one week really emphasized pipelines, as a case study of environmental health communications. Matt Smith generously visited the class again and mobilized students, providing an overview of pipeline developments and current advocacy opportunities. For a weekly assignment, I asked teams in the class to prepare and present examples of social media outreach for environmental health topics of their choice. With student permission, we incorporated Drew participation into CAPP posts, sharing photos of our meetings, and activities at outreach events in Madison, such as a "Green Fair" held at the library.

Since 2016, I have encountered much more literature concerned about the use of social media in advocacy. While academic public health literature continues to share examples of positive social medial outreach, my own feelings now are much more conflicted about the appropriateness of integrating social media into the classroom. The surveillance of social media use by advocacy groups [2, 52], and the misuse of social media by authoritarian governments [14] raise for me more arguments against social media use in the classroom than in favor. The recent book by Sarah Roberts, *Behind the Screen: Content Moderation in the Shadows of Social Media*, a case study of environmental injustice, discusses the exploitation of labor involved in content moderation for commercial enterprise that is intentionally disguised, despite the portrayal of social media companies as a free and open internet platforms [43].

Though my own views of social media have transformed significantly, leaning largely toward concern about the deleterious influence of social media on democracy, I try to remain open to mindful and useful digital communication strategies. I used and learned from the valuable text, website, and accompanying video documentaries by *New York Times* columnists Nicholas Kristof and Sheryl WuDunn [27]: *A Path Appears*. Kristof and WuDunn described a particularly effective Bill and Melinda Gates-funded organization called RESULTS. RESULTS offers a free introduction to advocacy courses, for individuals and for college classes, which I found valuable as a novice advocate-teacher [41].

RESULTS advocacy uses video conferencing, but generally focuses on traditional democratic engagement practices: civics education (how bills become laws), how to propose legislation, how to organize letter writing campaigns, and how to support network building to facilitate work on anti-poverty programs, locally and internationally. I have also found the work of Matthew Desmond, and the Eviction Lab that he supervises at Princeton University, a valuable model for CBL, and social science research generally: the housing eviction crisis in this US clearly pertains to environmental justice [13].

4.2.3 Community Mapping with the Sierra Club

Beyond applications to public health, opposition to the Pilgrim Pipeline presented opportunities for spatial analysis. In my earliest work as a volunteer with CAPP, I wanted to learn how I might support their work with my experience with Geographic Information Systems (GIS). The technical expert for the New Jersey chapter of the Sierra Club, Joe Testa, created and maintained a powerful web map (Fig. 4.1). The map shows the location of the proposed pipeline and bordering towns and communities that had passed resolutions against the pipeline (Coalition against the Pilgrim Pipeline (CAPP) [30]. For each location, the webmap provides a link to the resolution on each town's local government webpage. The map was on display at CAPP meetings, and used in CAPP canvassing. As a geographer, I loved the idea of a coalition that was truly united through this map.

Initially I developed some instructional materials to demonstrate how advocates could download, save, and edit the CAPP map for their own uses. It was a good opportunity for me to learn the tools available in the new Google My Maps features

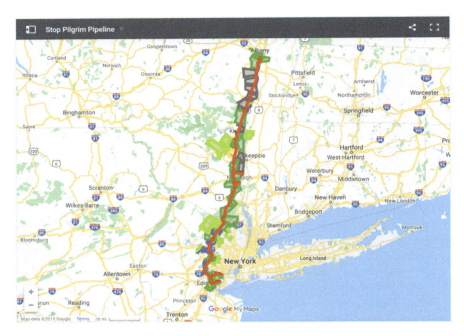

Fig. 4.1 Google my map of proposed Pilgrim Pipelines route and towns, available online [9]

[21], which are free to use. A CAPP subgroup organized around the idea of providing instructional materials for making community maps, and a woman with significant organizing experience, Ann Bastian, led our work. The tool allowed people in each township to zoom into their town, see the pipeline route, and use that example to identify areas important to them that might be affected by the route. By looking at the map, public meetings about the pipeline could focus on community assets that they valued and demonstrate potential disturbances and traffic problems that might be generated during pipeline construction.

I participated in reviewing some of the materials and guidelines for community mapping. I held a community mapping exercise at Drew to a small, but friendly audience, and we walked through the steps of map-making from the original map. Jerome Wagner, a 350.org organizer, was the official CAPP representative, and I also started to get to know Stephan Stocker, who chaired the Madison Environmental Commission for many years. A retired Vietnam vet, who will tell you about his miraculous survival of a grenade explosion, Stephan has always been generous with his time, and kindly mentored me on many occasions on the environmental landscape of Madison and New Jersey.

Using Google My Maps or other free online tools for sustainability mapping are straightforward applications to demonstrate to students or new users. It was a simple exercise to also replicate in my introductory Environment, Society and Sustainability class, as a case study in community mapping and exploring potential local pipeline impacts. We also collaborated in my classes to make a Drew sustainability map and map of locations with water refill stations, to help avoid unnecessary purchases of single-use disposable water bottles on campus.

4.2.4 Using Story Maps in GIS Coursework

While I enjoyed community mapping very much, and learned a great deal from the process, I wanted to help students develop the technical abilities to answer difficult questions related to controversial developments, such as a pipeline. Following my early work with community mapping, I became aware of new online tools to create visually engaging and professional maps called story maps [18]. After becoming better acquainted with the software, it became clear that a perfect example for teaching story maps would be the proposed Pilgrim Pipelines.

In fall 2016, I divided the GIS class into teams to explore different aspects of the pipeline that would affect communities: properties along the pipeline, wetlands, species-based habitats, cultural resources along the pipeline, sole source aquifers, historic properties, New Jersey schools, NJTransit rail intersections, category one streams (streams with a year-round flow and no tributaries), and the New Jersey Highlands Protection and Planning Area (Fig. 4.2). We created a distance buffer for the proposed pipeline route, then visualized and quantified valuable community and environmental resources that may be negatively affected by pipeline development.

Our story map was well received by CAPP. We presented our class work to the Chatham Environmental Commission (Chatham neighbors Madison), who shared our work on their website [16]. We also presented our work the following spring at the New Jersey DEP's annual GIS Poster Contest [6]. Our story map received

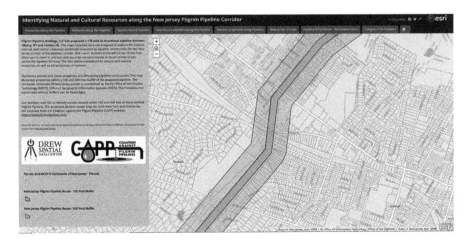

Fig. 4.2 Screenshot of Pilgrim Pipeline Story Map, available online [15]

second place. GIS professionals complimented the map, and many had never heard of the Pilgrim Pipeline. CAPP hired a consulting firm, Princeton Hydro, that used our stream data, which we extracted from the larger NJDEP data set, to look at the slope and drainage around the proposed pipeline near category one streams.

4.3 Discussion

The lasting takeaway of civic engagement for me has been to discredit the frequently stated notion and expectation of rugged individualism and loss of civic engagement, which I grew up with and that was described in Robert Putnam's *Bowling Alone* [39]. New Jersey politics is filled with extraordinary frustrations, but the capacity for community action and involvement that exists is remarkable. Now that non-profit, grassroots organizing is visible to me, I am able to see it everywhere. Organizations like the Association for New Jersey Environmental Commissions (ANJEC) and Sustainable Jersey are paving institutional pathways for civic engagement, particularly around environmental issues. Educators can take advantage of these pathways to integrate real-world participation into our courses.

As of 2019, both the Pilgrim Pipeline Holdings, LLC, and the CAPP websites remain online. The development process is stalled. Regular meetings were discontinued, though occasionally someone will post a pipeline-related article to the CAPP FACEBOOK page. Before pipeline debates died down, they did heat up. The Pilgrim Pipeline protest built on Dakota Access Pipeline energy. The Split Rock Sweet Water Prayer Camp highlighted long-neglected interests of the Ramapough Lenape Nation along the New Jersey–New York border, and received reporting from the *New York Times* [40]. I represented CAPP at Split Rock Sweet Water in Mahwah, New Jersey and found it a humbling experience. Bold, young men and women shared their experiences of peaceful resistance and honoring their elders and the earth. I was also impressed with the gentleness and quality of legal representation that the Eastern Environmental Law Center advocates provided. Concerns around the camp are ongoing [34].

Investigative reporter Derek Seidman traced Pilgrim funds to Koch Industries, Inc. [45] and Governor Phil Murphy's administration [46], leading some to feel that the pipeline proposal could resurface. Though President Trump's administration has led to rollbacks of many environmental regulations and norms, the replacement of Governor Chris Christie in New Jersey with Governor Phil Murphy seems to have had some positive effects on state environmental regulation. New Jersey DEP's Commissioner, Bob Martin, was replaced by Catherine McCabe [12], an environmental attorney and EPA administrator, and Debbie Mans, an NY/NJ Baykeeper became the deputy commissioner [20]. Locally, Chatham and other towns have explored passing zoning ordinances to limit the development of new pipelines (Township of Chatham, New Jersey [48]. Several nearby towns formed a coalition for gathering legal fees to defend towns from pipeline construction, in the event that pipeline development moves forward in the future [24].

Renata Schiavo has argued that even when activists are unsuccessful in achieving their goals, that the process itself can be a success: the building of community relationships and connections, the education and outreach that takes place, and the road traveled down the long arc toward justice are all worthy. In other words, losing can be winning [44]. Robert Bullard echoed this sentiment when he described the legal ruling that found the waste company 'not guilty' for siting dumps in poor, black communities in *Bean v. Southwest State Management Corp.* (1979): the case brought to light and called attention to significant environmental inequalities that had not been widely noticed before. He argues: "it remains a watershed case because it was the first instance in U.S. history in which Americans challenged environmental racism using civil rights law" [4].

Conversely, winning can be losing. The settlements received on behalf of those who have experienced environmental discrimination often do not begin to approximate adequate compensation for the losses experienced, often over long periods of time, and involving the health of loved ones and the health of entire communities. The process of the Pilgrim Pipeline proposal brought together communities, a wide variety of activists, and demonstrated our capacity to work together to achieve a common goal. The absence of new information from Pilgrim Pipeline Holdings, LLC brings loss of interest, without necessarily a transfer to other, related work. The extensive place-based development proposal brought people together across state-lines. At the moment, it is a win with some loss.

Nevertheless, along this journey, becoming a more humble educator, and a more active community member, has been a persistent benefit of CBL pedagogy. I have deeply benefited from the realization that valuable teachers are easily within reach and interested in working with me and the young people I am privileged to work with everyday. There is a deep, and for me surprising, willingness of others to help one another. I look forward to working with and learning from my community in the future, and I take joy in seeing my students become active participants in our shared community.

4.4 Conclusion

In my experience, bringing pipeline examples into college classes creates a learning environment that values local knowledge. Local knowledge is found through civic engagement, learning from and listening to neighbors, and exploring the places we live, work, and play in the context of energy development. The process of coming to understand the world around us, writ-small, helped to expand our capacity to read, interpret, and understand other case studies, taking place in other locations. Learning about CAPP and the Pilgrim Pipeline elevated the importance of local news sources and local leaders as experts.

CBL pedagogy can serve as a relatable place-based case study approach that connects the theory and practice of public health and environmental studies. Current research has found that case studies "can increase student motivation to participate in class activities, promoting learning and increasing performance on assessments" [7]. Additionally, exploring a case study in detail "affords students the opportunities to engage with ethical and societal issues related to their disciplines, as well as facilitating interdisciplinary learning" [7]. I found this to be the case in my classes. But, access to this approach, for me, required institutional support from the university and from organizations like Campus Compact.

A question that might be asked of ESS faculty is: how can environmental studies and sciences support communities facing environmental injustices brought on by pipeline development? In our case, it was the ability to showcase our abilities and talents, simply to visualize our area and the impact caused by pipeline development. Our presence and witness at meetings served to support education and awareness of development projects. In some ways, university involvement may provide legitimacy to opponents in pipeline disputes. But, I think it is important to recognize a growing portion of the US population that distrusts the work of universities [35, 37].

My hope is that by sharing my story of how I used pipeline disputes in the classroom will provide teacher-scholars with ideas and information about how civic engagement can enrich classroom materials. For community organizers, I hope sharing the perceptions of my experience as a scholar involved in advocacy work, and the type of balancing and decision-making that I considered in my participation is useful to you. There is a misconception that professors should know everything, but my experience has taught me that I am mainly a learner, alongside my students, in new situations. For researchers, I would encourage and advise work as an activist, because through activism you have an opportunity to learn from many, varied experts, and to become familiar with the values and ideas that communities hold in high esteem.

I think the work expressed here relates to pipelines elsewhere. However, civic engagement pertains not only to pipelines, but to other types of development projects. Ongoing Superfund site management, brownfield development, everyday planning work (pedestrian audits, bicycle audits), and everyday public health education and outreach (e-cigarettes) share many intersections with course topics in environmental studies and sciences. My own work continues to be actively inspired by extraordinary scholar-activists: Robert Bullard and Beverley Wright [5], Matthew Desmond and his work with the Eviction Lab at Princeton University [19], Alex de Waal's writings with the World Peace Foundation [11], Naomi Klein's coverage of agroecology education in Puerto Rico [26], Umair Muhammad's [33] work and writing in relation to Jane-Finch Actions Against Poverty in Toronto, and scholarship on place-based learning that shares stories from other campuses [53].

References

1. Baptista A (2019). Local policies for environmental justice: a national scan. Tishman environment and design center. Retrieved from https://static1.squarespace.com/static/5d14dab43 967cc000179f3d2/t/5d5c4bd0e1d5150001a5a919/1566329811163/NRDC_FinalReport_04. 15.2019.pdf

2. Brown A, Parrish W, Speri A (2017, May 27) Leaked documents reveal counterterrorism tactics used at Standing Rock to "defeat pipeline insurgencies." The intercept. Retrieved from https://theintercept.com/2017/05/27/leaked-documents-reveal-security-firms-counterterrorism-tactics-at-standing-rock-to-defeat-pipeline-insurgencies/
3. Bullard RD (1993) Confronting environmental racism: voices from the grassroots. South End Press, Boston
4. Bullard RD (2005) The quest for environmental justice: human rights and the politics of pollution. Sierra Club Books, San Francisco
5. Bullard RD, Wright B (2012) The wrong complexion for protection: how the government response to disaster endangers African-American communities. NYU Press, New York
6. Bureau of GIS, Department of Environmental Protection, State of New Jersey (2017) 30th Annual NJDEP Mapping Contest Winners. 30th Annual NJDEP Mapping Contest Winners. Retrieved from https://www.nj.gov/dep/gis/mapcon30.html
7. Burns W (2017) The case for case studies in confronting environmental issues. Case Studies in the Environment 1(1): 1–4. https://doi.org/10.1525/cse.2017.sc.burns01
8. Campus Compact (2019) Campus compact: mission & vision. campus compact. Retrieved from https://compact.org/who-we-are/mission-and-vision/
9. Coalition Against the Pilgrim Pipeline (CAPP) (2014, October 30) Coalition against the Pilgrim Pipeline (CAPP): Maps. Retrieved from https://stoppilgrimpipeline.com/maps/
10. Commission for Racial Justice, United Church of Christ. (1987) Toxic wastes and race in the United States: A national report on the racial and socio-economic characteristics of communities with hazardous waste sites. Retrieved from https://omeka.middlebury.edu/fyg/items/show/316
11. de Waal A (2017). Mass starvation: the history and future of famine. John Wiley & Sons, Hoboken
12. Department of Environmental Protection, Office of the Commissioner (2018, June 20) New Jersey Department of Environmental Protection-Commissioner. Retrieved from https://www.nj.gov/dep/commissioner/
13. Desmond M (2016) Evicted: Poverty and profit in the American city. Crown Publishers, New York
14. Dobson WJ (2012) The dictator's learning curve: inside the global battle for democracy. Knopf Doubleday Publishing Group, New York
15. Drew University (2019) Drew University: civic engagement. Retrieved from http://www.drew.edu/civic-engagement
16. Environmental Commission, Borough of Chatham, New Jersey (2019) Borough of Chatham: Pilgrim Pipeline Information. Retrieved from https://www.chathamborough.org/chatham/Pipeline/
17. EPA Alumni Association (2018, November 23) EPA Alumni association: alumni in the classroom. Retrieved from https://www.epaalumni.org/classroom/
18. ESRI (2019) ArcGIS StoryMaps: Storytelling that resonates. Retrieved from https://storymaps.arcgis.com
19. Eviction Lab (2018) The eviction lab. Retrieved from https://evictionlab.org/
20. Fallon S (2018, February 6) Appointment of veteran environmentalist marks shift at New Jersey DEP. Daily Record. Retrieved from https://www.northjersey.com/story/news/environment/2018/02/06/appointment-veteran-environmentalist-marks-shift-new-jersey-dep/309138002/
21. Google (2019) My Maps Help. Retrieved from https://support.google.com/mymaps#topic=3188329
22. Hackensack Riverkeeper (2019) Eco-cruises—Hackensack Riverkeeper. Retrieved from https://www.hackensackriverkeeper.org/activities-and-events/eco-cruises/
23. Javorsky N (2019, May 7) Which cities and states have concrete strategies for environmental justice? CityLab. Retrieved from https://www.citylab.com/equity/2019/05/environmental-justice-racism-zoning-land-use-baltimore-nyc/588793/
24. Kadosh M (2019, May 10) Millburn may join consortium opposing pilgrim pipeline. Daily Record. Retrieved from https://www.northjersey.com/story/news/essex/millburn-short-hills/

2018/05/10/millburn-nj-may-pay-into-north-jersey-group-opposing-pilgrim-pipeline/569969002/
25. Keill L (2015, February 26) Pilgrim pipeline issue draws major crowd in Madison. TAPintoMadison. Retrieved from https://www.tapinto.net/towns/madison/articles/pilgrim-pipeline-issue-draws-major-crowd-in-madis
26. Klein N (2018) The battle for paradise: Puerto Rico takes on the disaster capitalists. Haymarket Books, Chicago
27. Kristof ND, WuDunn S (2015) A path appears: transforming lives, creating opportunity. Knopf Doubleday Publishing Group, New York
28. Mennis JL, Jordan L (2005) The distribution of environmental equity: exploring spatial nonstationarity in multivariate models of air toxic releases. Annals of the Association of American Geographers 95(2): 249–268. https://doi.org/10.1111/j.1467-8306.2005.00459.x
29. Miller-Travis V (2018, October 16) Environmental justice in the 21st Century: threats and opportunities (Part 3). American University College of Law. Retrieved from https://www.youtube.com/watch?v=I87C9ki1SYE&feature=youtu.be
30. Millsaps K (2014 September 8) We need your help to fight this pipeline! Coalition Against Pilgrim Pipeline (CAPP) Retrieved from https://stoppilgrimpipeline.com/2014/09/08/we-need-your-help-to-fight-this-pipeline/
31. Milman O (2019, May 21) Revealed: 1.6 m Americans live near the most polluting incinerators in the US. The Guardian. Retrieved from https://www.theguardian.com/environment/2019/may/21/us-pollution-incinerators-waste-burning-plants-report
32. Mouawad J (2014, February 24) Bakken crude, rolling through Albany. New York Times. Retrieved from https://www.nytimes.com/2014/02/28/business/energy-environment/bakkan-crude-rolling-through-albany.html?_r=0
33. Muhammad U (2017) Confronting injustice: social activism in the age of individualism. Haymarket Books, Chicago
34. Nir SM (2019, April 22) Native Americans find surprising ally in N.J. Fight: Trump Administration. The New York Times. Retrieved from https://www.nytimes.com/2019/04/22/nyregion/ramapough-lenape-indians-mahwah-nj.html
35. Pew Research Center (2017, July 10) Sharp partisan divisions in views of National Institutions. Pew Research Center for the People and the Press. Retrieved from https://www.people-press.org/2017/07/10/sharp-partisan-divisions-in-views-of-national-institutions/
36. Pilgrim Pipeline Holdings, LLC. (2014). Project description. Pilgrim Pipeline Holdings, LLC. Retrieved from http://pilgrimpipeline.com/project-description/
37. Pinsker J (2019, August 21) Republicans changes their mind about higher education really quickly. The Atlantic. Retrieved from https://www.theatlantic.com/education/archive/2019/08/republicans-conservatives-college/596497/
38. Pipelines International (2014, April 10) Pilgrim proposes new pipeline to New York Harbor. Pipelines International. Retrieved from https://www.pipelinesinternational.com/2014/04/10/pilgrim-proposes-new-pipeline-to-new-york-harbor/
39. Putnam RD (2001) Bowling alone. Simon and Schuster, New York
40. Remnick N (2017, April 14) The Ramapoughs vs. the world. The New York Times. Retrieved from https://www.nytimes.com/2017/04/14/nyregion/ramapough-tribe-fights-pipeline.html
41. Results (2019) Homepage. Retrieved from https://results.org/
42. Riverkeeper (2016). Crude oil transport. Riverkeeper. Retrieved from https://www.riverkeeper.org/campaigns/river-ecology/crude-oil-transport/
43. Roberts ST (2019) Behind the screen: content moderation in the shadows of social media. Yale University Press, New Haven
44. Schiavo R (2013) Health communication: from theory to practice. John Wiley & Sons, Incorporated, Hoboken
45. Seidman D (2017, March 8) The power behind the pipelines: pilgrim pipeline. Public accountability initiative. Retrieved from https://public-accountability.org/report/the-power-behind-the-pipelines-pilgrim-pipeline/

46. Seidman D (2018, March 22) If the Pilgrim Pipeline is dead, why is it still lobbying in New Jersey? Little sis: eyes on the ties. Retrieved from https://news.littlesis.org/2018/03/22/if-the-pilgrim-pipeline-is-dead-why-is-it-still-lobbying-in-new-jersey/
47. Spatial Data Center, Drew University (2019) Identifying natural and cultural resources along the New Jersey Pilgrim Pipeline Corridor Retrieved from http://tinyurl.com/capp-gis
48. Township of Chatham, New Jersey (2018, January 3) Pilgrim Pipeline. Retrieved from https://www.chathamtownship-nj.gov/pilgrim-pipeline
49. US EPA (2019, May 24) TRI university challenge [Announcements and Schedules]. US EPA. Retrieved from https://www.epa.gov/toxics-release-inventory-tri-program/tri-university-challenge
50. US EPA O (2013, July 24) Emergency Planning and Community Right-to-Know Act (EPCRA) [Collections and Lists] US EPA. https://www.epa.gov/epcra
51. Waldman S (2014, March 11) Company explores Albany-New Jersey crude pipeline. Politico PRO. Retrieved from https://www.politico.com/states/new-york/albany/story/2014/03/company-explores-albany-new-jersey-crude-pipeline-011526
52. Wilson J, Parrish W (2019, August 8) Revealed: FBI and police monitoring Oregon anti-pipeline activists. The Guardian. Retrieved from https://www.theguardian.com/us-news/2019/aug/08/fbi-oregon-anti-pipeline-jordan-cove-activists
53. Yamamura EK, Koth K (2018) Place-based community engagement in higher education: a strategy to transform universities and communities. Stylus Publishing, LLC, Sterling

Part II
Tools and Methods for Teaching Pipeline Controversies

Chapter 5
The Stop PennEast Pipeline Fieldwork Project: Teaching Students to Apply Fieldwork Methods to Studying a Natural Gas Pipeline Opposition Movement

Michael J. Brogan

Abstract The Stop PennEast fieldwork project focuses on teaching students how to apply social fieldwork methods to studying an opposition natural gas pipeline movement, located in northeastern PA and northwestern NJ. The fieldwork was connected to an engaged learning assignment offered as part of an Environmental Politics class at Rider University from 2014 to 2018. Course fieldwork was designed to connect data collection and analysis efforts with existing literature covered in class on social movements, environmental networks, and information diffusion between network actors. Work connected to the project included conducting field surveys, canvassing, participant-observations of movement meetings, a survey of state legislators in NJ and PA (in collaboration with the Delaware Riverkeeper Network, an environmental advocacy not for profit), and content analysis of public testimony and social media exchanges. This chapter focuses on having students conduct field surveys of Stop PennEast meeting attendees at five public meetings from November 2014 to January 2015.

Keywords Fieldwork methods · Environmental politics · PennEast pipeline · Engaged learning · Opposition movement · Natural gas

5.1 Introduction

This chapter focuses on the Stop PennEast fieldwork project. This pedagogical project engaged students in applying social science fieldwork methods to study an opposition natural gas pipeline movement, located in northeastern PA and northwestern NJ. The fieldwork was connected to an engaged learning assignment offered as part of an Environmental Politics class at Rider University from 2014 to 2018. Course fieldwork was designed to connect data collection and analysis efforts course content on social movements, environmental networks and information diffusion

---

M. J. Brogan (✉)
Department of Political Science, Rider University, Lawrenceville, NJ, USA
e-mail: mbrogan@rider.edu

© Springer Nature Switzerland AG 2021
V. Banschbach and J. L. Rich (eds.), *Pipeline Pedagogy: Teaching About Energy and Environmental Justice Contestations*, AESS Interdisciplinary Environmental Studies and Sciences Series,
https://doi.org/10.1007/978-3-030-65979-0_5

between network actors. Students completed two main fieldwork assignments: (1) A field survey of Stop PennEast meeting attendees at a PennEast Open House forum in November 2014; (2) A limited participant-observation assignment at a Federal Energy Regulatory Commission (FERC) Scoping Meeting in March 2015.

I present tools and methods for individuals who not only want to develop rich data sets and ideas from the field, but to also connect these efforts with their teaching. Though fieldwork is essential to learning more about the context of political and social interactions and problems, it is often an overlooked skill set. This is evident in a systemic lack of training for undergraduate and graduate students other than an "appreciative nod" to Richard Fenno's idea of "soaking and poking" [1]. Teaching students about fieldwork methods, and applying them, not only provides an engaged experience for students but also substantiates specific lessons as to how researchers and practitioners can further understand the context of deeply complex issues. Much of the literature on conducting fieldwork, tends to focus on overly generalized themes and logistics [2]. By applying fieldwork to studying opposition to a natural gas pipeline, students learn about the challenges and ethical dilemmas of interacting with people who are suspicious of the motives of the researcher and the direction of the research generated from the project [12].

I outline a description of two assignments related to the Stop PennEast project. The first is an information diffusion assignment in which students conducted field surveys of movement members and participant-observations of public meetings. The second involves students acting as participant observers at a FERC Scoping Meeting. I present the student learning goals of the assignment, methods used to study the opposition, and a summary of the project's findings. Finally, the chapter provides an assessment of the project and lessons learned from the project for practitioners and academics seeking to conduct an in-depth study of environmental movements.

5.2 A Brief Background of the PennEast Pipeline Project

The proposed PennEast Pipeline would bring natural gas to customers in Pennsylvania and New Jersey. At a cost of approximately $1 billion, this new 114-mile, 36-inch diameter pipeline promises to deliver approximately 1 billion cubic feet of natural gas per day. The pipeline would originate in Dallas, Luzerne County, in northeastern Pennsylvania, and terminate at the Transco pipeline interconnection near Pennington, Mercer County, New Jersey [8].

Primary issues for opposition groups on PennEast include: (1) Environmental issues–loss of habitat, pollution, damage to existing waterways, over-usage of water (hydrostatic testing), and hydraulic fracturing. (2) Land use–loss of property due to threat of eminent domain; unfair compensation of Right Of Way (ROW), and (3) Safety—PennEast would be a 36-inch natural gas pipeline where opposition concerns focused on the blast radius (of 900ft) and from thermal radiation from a blast (~1800ft); the estimated danger zone would be roughly a 0.65mi danger zone from point of impact. Concern over safety would be based upon flammability and risk for the general area of a pipeline due to the probability of rupture and release of the fuel which potential could cause a fire and explosion [9].

Since the public announcement of the PennEast pipeline, the project can be summed up by a continuing theme: delay. In April 2016 the Federal Energy Regulatory Commission (FERC), which oversees permitting for the pipeline, told PennEast the agency would extend the amount of time they are taking until December 2016, rather than the original target of August, to complete their environmental review. Yet, the company experienced further delays in achieving final environmental review by December 2016 and had to push the date to February 2017. By that time, the New Jersey Department of Environmental Protection (NJDEP) and PA DEP had to issue 401 water permit. Under the US Clean Water Act, 401 permits are issued by states where any activity that results in the discharge of water in the jurisdiction is in compliance with the regulation [10]. Issuing this certification was particularly problematic in NJ because of lack of surveys in NJ, where only 35% of properties were completed. Another delay was announced in April 2017 by FERC saying that the final environmental review has not happened. In December 2018, PennEast received assistance in finishing environmental review when the company won a court case in US Middle District Court of PA. The case ruled in favor of the company being able to declare eminent domain of private properties. The ruling allowed the company to complete land surveys and environmental assessment on properties that were inaccessible prior to the decision because of landowner resistance [3]. The company announced they were pleased with the ruling and planned to begin construction by mid-2019. In August 2019 the company reapplied for permits from NJDEP, which were originally denied because of the lack of environmental surveys completed.

As of November 2019, the company experienced another setback. The U.S. Third Circuit Court's opinion ruled that a private entity cannot condemn state-owned property. Much of the proposed route consists of state lands thus making it difficult for the Company to proceed. This is a victory for those opposed to the pipeline; however, the project has not been completely stopped. The Company has decided to appeal the Circuit Court's ruling to the US Supreme Court. PennEast could also persuade the Federal government to do condemnation of state lands, assuming the Federal Energy Regulation Commission (FERC) would carry this out. In any event, the original PennEast proposed route and calendar illustrate how long the project has been delayed. The company first proposed the route in August 2014, stated the permits would be approved by 2016, and vowed for construction and completion of the project by 2017. PennEast is now three years behind its initial completion projections and cannot move forward until either its legal disputes are resolved or if the federal government provides support by condemning state lands on the company's behalf [6].

5.3 Student Learning Goals

Student Learning goals are connected to an upper level undergraduate course entitled Environmental Politics. The course examines how policymakers deal with the political challenges of unsustainable resource consumption, which is a primary determinant of environmental problems such as climate change, adverse health effects, and biodiversity loss. It looks at the policy and political process related to environmental problems. Per this fieldwork assignment, students focused on the process component related to this issue. The course is organized around two major learning modes: lecture and seminar formats.

The field project guides students to complete an experiential activity, a process of learning by doing outside of the classroom. Learning goals were designed in order to prepare students to conduct research in the "field." The broader benefits of using the proposed PennEast pipeline project were that the students connected with the local community, gained the personal experience of grassroots political action, experienced a sense of camaraderie (as we traveled together into the field to complete the project) and ownership of their research. The PennEast pipeline fieldwork project was designed around the following learning goals:

1. Students become able to describe the background, context, and actors involved in the project
2. Students are trained to apply field research techniques in order to begin to create a social network or loosely affiliated actors around the problem.
3. Students are able to analyze data from the field research and present results
4. Students connect and contextualize the problems of collective vs individual needs in the setting and distribution of outcomes within this case.

By design, course learning goals closely follow logistical requirements of the project (e.g. timing to do fieldwork throughout the term with actions and events coordinated by outside groups). In addition, learning goals are scaffolded, meaning prior goals build on to the next set of goals. Here, I will summarize the work linked to each learning goal. Later in the chapter, I fully describe each assignment.

The first learning goal focuses on students conducting background research on the project. Students were required to present and analyze the benefits and limits of the proposed pipeline as well as the broader structure of natural gas energy infrastructure, how it is regulated, the role of industry and organizations/groups that are for and against pipeline expansion. The assignment informed students of the context in which they would conduct their fieldwork. Students developed effective research skills and communication skills related to the issue.

For the second learning goal, students are trained in fieldwork techniques of interviewing, participant-observation. First, students complete a training session per Institutional Review Board guidelines. The instructor had students do practice interview sessions of other students on campus. Also, students became participant observers at student group meetings. Students also developed skills in conducting professional field interviews and in observing and documenting events and participant behavior.

The third learning goal required that students collect, manage, report, and analyze their findings. This was done informally, at the beginning of the class meeting right after students went into the field, students gave an after-action report of their observations and findings. At the end of the term, students formally presented their observations, reflections on the process and recommendations for next steps. Students effectively developed skills in collecting and interpreting findings.

Finally, students were asked to write brief reflection papers that both described what they learned from the project but also how their efforts fit into the broader debate over pipeline expansion. Students were asked to evaluate their impact on the people they met and how the project impacted their perspective regarding whether there should be a continued expansion of natural gas pipelines. Students developed skills to synthesize their experience with what they learned in the field with their coursework.

5.4 First Assignment: Attending a Public Meeting and Completing a Field Survey

For the first assignment students administered a Pipeline Connections survey on November 13, 2014 at the PennEast Open House at South Hunterdon High School. Students interviewed and collected data from 56 attendees at the event. The survey instrument can be found in the Appendix. All completed survey forms were submitted to the instructor at the end of the forum and overall aggregate data was provided shortly thereafter. Students made observations about the event (using their interview forms to guide them). Face-to-face interviews were done by students who used a structured questionnaire that was administered to all people who were leaving the event (interviews happened outside of the school, respondents chose whether to complete the survey) [5]. In addition, students answered assigned questions from background readings for the week, from Meyer [7] and Gerlach [4], on environmental movements and protests before the event and then the class discussed their responses and observations in class after completing the field work. The survey and discussion questions are included in the appendix.

5.5 Second Assignment: Limited Participant-Observations of a PennEast FERC SCOPING

For this assignment, students worked in pairs to collect and interpret information about a Federal Energy Regulatory Commission (FERC) Scoping meeting through a limited participant-observation process [11]. FERC scoping meetings are public meetings where staff collect and identify relevant issues of projects that require certification under the National Environmental Policy Act (NEPA). The process involves public participation and comments from affected stakeholders who can suggest alternatives, constraints, and provide information on environmental features in the project area.

Scoping meetings are an ideal venue in which to conduct limited participant-observation fieldwork. Limited participant-observation is a qualitative research method with the goal of better understanding a population of research interest by direct observation and participation by researchers. What researchers are doing is attempting to learn about those directly involved (e.g., an "insider" in a social movement) while remaining outside of the group (Mach 2005). In the case of this project, students participated in activities of the Stop PennEast opposition movement being observed but did not become members of the movement [5].

The assignment placed particular emphasis on how students assessed the meeting. To do this, they needed to gather enough evidence to analyze the event by recording all observations into their notes and share them with me before the next class meeting. Students were given the following prompt:

To complete the Limited Participant-Observation assignment, students notes addressed the following:

- Describe the scene, paying attention to all sensory perception. Draw a map of the setting, indicating the position and movement of persons. Who is present? Who is absent?
- Look for the structure of the situation: are the participants differentiated from each other, as, e.g., leaders and followers, or those with more or less status? Is status differentiation or equality represented in dress, behavior, symbolic markers, differing prerogatives? How do people interact with each other?
- Are there any elements of power, either formal or informal, in what you observe? How do you interpret the meaning and effect of how power is distributed among members at the meeting?
- What appears to be the formal, or informal, rules that guide this event or activity? Is there any mechanism for correcting a distortion or a mistake, any formal authority? Do people seem to follow the rules, explicit or tacit, or do they bend them?
- Is the event characterized more by order and agreement or conflict and disorder?
- Ask two people why they came to the meeting (listen for specifics), how they define the issue (what's the real problem in their words?), what are the reasons for others who came that might support their position and those who may not support their position (can they empathize with those they disagree with)?
- Do all participants seem to be deriving the same benefits or satisfactions from participation? Do they have means of communicating positive or negative judgments about the situation?
- Are there coordinated actions or modes of self-presentation that seem characteristic of this event? Do the participants seem aware of them? What purpose do they serve?
- What shared values or assumptions are reinforced (or contested) through this event?

Students submitted the following: (1) Raw field notes (2) An after-action elaboration (1-2 pages) that included a summary and description of the Limited Participant-Observation (scene-setting and highlights), analysis and interpretation of the event, and self-reflection (did they participate and to what extent did they observe; what could they have done better)?

5.6 Third Assignment: Analyzing Findings

Students conducted 56 field interviews at the PennEast Open House held at South Hunterdon High School located at 301 Mt Airy-Harbourton Rd, Lambertville, NJ on November 13, 2014. The most common way individuals heard about the Open House meeting was email (19%), newspaper (16%), word of mouth (12.5%), social network (9%), and via a flyer (7.5%). About 44% of respondents were contacted prior to the meeting. Of those who were contacted the most common contact came from neighbors (36%), pipeline company and township (9% each), and about 7% from environmental groups. On average, respondents spoke to about 33 people about the proposed pipeline. On a scale of 1 to 5 (lower scores less knowledgeable and higher scores more knowledgeable), the average self-reported knowledge level about the PennEast pipeline was 3.8.

For Limited Participant-Observation findings, students noted that around 200–400 people attended the FERC Scoping meeting held at 2015 at the West Trenton Ballroom at 40 W. Upper Ferry Road West Trenton, New Jersey on February 25, 2015. Student observations from the event include:

- "The meeting had not started when I arrived, but the building was crowded. Upon realizing there was no more parking, I had the feeling it would be a mad house. There were a lot of people standing alongside the walls and in the rear of the room prior to the start of the meeting. There were heavy equipment operators and various other employees of the Penn East pipeline project and New Jersey and Pennsylvania residents present at the meeting. Those in support of the pipeline were in the rear of the room while those in opposition were either standing on the sides, seated in chairs or standing towards the front of the room. I noticed people with signs expressing their discontent with the pipeline."
- "I noted a sense of tension amongst the FERC board members and with the audience, which was overwhelmingly, opposed to the pipeline."
- "Those who spoke during the meeting shared facts, statistics, and research study finds, but most importantly they shared the stories of their families' history and how it would affect future generations to come."
- "Two of the most memorable speakers at the FERC scoping meeting were brother and sister Jaycee and Kaia, two children around the ages 12 and 9 years old. They pleaded for FERC to think about future generations, generations whom will not get to enjoy farmland, fresh water, and the wildlife. Kaia chose to mention that there is already a pipeline in her backyard, and was fearful and strongly against

the implementation of a second one. She offered the board members the question of "why are they building new pipelines, as opposed to correcting the ones already in place?"
- "Even though I have been voting for 3 years that was my first firsthand experience of democracy. Over three quarters of the room did not want the pipeline. the other quarter was a mix of impartial people and those steam workers bussed in so that it would show how many lives could be affected by "jobs" that would be created by this so-called public works project."
- "The other two gentlemen I interviewed refused to give their names (one was a Union shop leader and the other a union worker local 825) and at first, they won't talk to me, but when I found out they were union workers, I mentioned that my father is a teamster and that I was a veteran. After hearing this, they were willing to talk to me. For them, the pipeline is about jobs and security and they resent anyone or anything that would prevent them from supporting their families. They felt that the environmentalist and activist were more concern about what may happen than people trying to make a living."
- "The event was, without a doubt, characterized by conflict and disagreement. Some environmentalists, who went up to speak, were booed by the laborers in the back of the room while they were cheered on by fellow environmentalists. Also, the environmentalists would stand up and cheer and the laborers would yell at them to sit down. This would happen at the end of almost every other speech. Beyond this, these two groups hardly interacted with each other. There was no attempt to debate openly. In addition to being characterized by conflict, the event was also very disorderly."

5.7 Lessons Learned

Overall, the lessons learned stem from final student reflection papers on the project as well as from the instructor's observations of conducting it. I present instructor observations, student feedback, and ways in which this project can be implemented in other settings. I conclude by highlighting overall themes from student post-course feedback.

For instructors, the biggest challenge is time. The project requires extensive course planning in terms of content and logistics. Planning meeting dates and activities should be done about one to two months before the term. A way in which to make planning easier is to reach out to outside groups who are working on a proposed pipeline project. These groups should provide logistical support in terms of providing a tentative calendar of planned actions, hearings and meetings. It is important to gain trust of these groups by disclosing what you plan to do and to share your aggregate findings with them if requested. To do so requires instructors to attend planning meetings prior to the term and participate in group phone calls or video chat sessions.

To ensure student success, doing preliminary outreach makes sure students are going to be successful at an event, action, or hearing with both folks inside and outside of the network. By doing this, it also ensures there will be people at the meeting who are already open to helping students complete their assignments. This is particularly important when doing the two assignments because it requires students to connect with individuals from outside of the university. At meetings, outside individuals, who include local residents and impacted community members, and organizers have been very helpful in making an announcement that acknowledges to their members that students from Rider University will be conducting an assignment at a particular event. The success of this project came from building extensive relationships with professional environmental organizations as well as with citizen groups from differing locations along the proposed pipeline.

5.8 Student Reflections and Themes from Completing Coursework and Fieldwork

Student reflections of the project also provide important lessons. First, students need to be made aware of the time commitment outside of normal class meetings. Students have many demands on their time so allowing a flexible calendar and multiple ways to earn credit increases likelihood of successful outcomes. Additional options given to students to complete the assignment were based on being able to write comments to public officials on the issue, get signatures for online petitions, canvassing for environmental organizations and impacted community groups, and attend other meetings or workshops.

Student reflections indicated the following themes related to collective action and civic engagement, student solidarity, enlightening students' perspectives on environmental problem and politics, and a call to action. The themes come from students' final reflection papers detailing what they learned and how the class impacted them. Overall students were very positive on the benefits of the class and the fieldwork assignments. Below I have provided summaries of students' comments on each theme:

5.8.1 Collective Action and Civic Engagement

A student noted the class project raised her awareness of collective action and civil discourse on the issue. She writes "[b]efore attending the meeting, I was aware of the pipeline and what it was hoping to accomplish. However, after seeing everyone speak about the issue, it was clear to see why the pipeline would destroy more lives than provides benefits and why people are against it."

5.8.2 Student Solidarity

Another student pointed out how the project helped bond students together: "Something I found very interesting was how everyone in our class had different political views, interest in fields of ecological development and methods to persuade others but we all worked together to spread our ideas and learn more about each other."

5.8.3 Enlightening Student Perspectives on Environmental Issues and Politics

In terms of engagement a student reflects that though getting involved in civic activism was "not something" she 'had ever done or planned to do." By the end of the term, participating in and observing environmental activism first hand showed her "how easy it is to get involved politically" and she "enjoyed each and every experience" she "engaged in this past semester." She writes in her self-reflection that "I learned so much from participating in politics throughout the semester and I hope everyone gets to experience what I did during this class." She concludes that at the start of the term she was not interested or planned to engage the political process but after the class writes "I would recommend this class to many people and I think I will be involved politically in the future because of my experiences in this class. Thank you!"

5.8.4 A Call to Action

Following on the theme about how doing field work helped trigger students to become motivated in engaging the political process, another student writes "taking action behind the scenes and in the classroom are important and necessary to spread education and start the sparks of change. However, that alone is not enough. To make a real impact, we have to get out of the classroom and engage with real influencers. The political process is not passive, and scientific knowledge is not found or spread by sitting around and waiting. For me, the biggest takeaway from this course that will stay with me for the rest of my life: if we want to spark change, we are the ones who need to get out there and start the fire."

As of the writing of this chapter, the PennEast pipeline project has been put on hold due to the US Circuit Court rejecting the company's appeal on being able to condemn state-owned lands. PennEast put out a press release stating it will take its appeal to the US Supreme Court. It is also likely the company will lobby FERC to ask for the federal government to condemn state-owned lands in order to build the pipeline. Though this is a tentative victory for those opposed to the project, it does show students about how civic activism and student involvement in working with the opposition movement can have a real, positive, impact on the environment and communities. The current state of the project and students' reactions to it has reinforced that student work in the field helped contribute to the movement and that engaging the political process is easier than previously considered.

Overall, I cannot emphasize enough the importance of student fieldwork. The current outcome of the proposed PennEast pipeline project highlights the meaningful impact of student fieldwork. Students have first-hand experience studying and participating in social change. I mentioned that fieldwork is often overlooked in teaching undergraduates, and setting up the PennEast pipeline project is logistically challenging. Yet, the overall skill set students developed over the course of a term, coupled with students seeing that their work made a difference, provides an invaluable lesson to anyone willing to take on this challenge to their teaching. In terms of student engagement, solidarity, learning and democracy, fieldwork is essential to empowering our next generation of leaders.

Appendix

PennEast Pipeline Survey

Introduction: "Hello, my name is [insert your name] and I am a student at Rider University. I would like to ask you a few questions regarding your feelings on the proposed PennEast pipeline project, reasons for coming out to today's meeting and how you have been able to stay connected to the issue. Your opinion is important to use to identify ways to better communicate to the public about this issue and ways to help identify all of the various groups who have been working on the issue. The survey should take no longer than 5–7 min to complete. Your identity will remain confidential and results will only be reported in the aggregate form."

1. Why did you come to today's meeting?

    - To support the PennEast pipeline
    - To oppose the PennEast pipeline
    - Other: (If "other" could you tell me your reason)

If "oppose" to question 1, then ask the following:

1. In defining your primary concerns with the proposed PennEast pipeline route, do you agree or disagree the proposal would:

    - Be bad for the environment (**A/D/NS**)
    - Cause safety issues (**A/D/NS**)
    - Increase health problems associated with pollution (**A/D/NS**)
    - Lower property values (**A/D/NS**)
    - Loss of property use (**A/D/NS**)

- Not being compensated by the company for the use of land in town to build the pipeline route (**A/D/NS**)
- Other: (If "other" could you tell me your concern(s))

If "support" to question 1, then ask the following:

2. In defining the primary benefits with the proposed PennEast pipeline route, do you agree or disagree the proposal would:
   - Create jobs (**A/D/NS**)
   - Provide property tax relief (**A/D/NS**)
   - Lower energy prices (**A/D/NS**)
   - Boost the local economy (**A/D/NS**)
   - Reduce greenhouse gas emissions (**A/D/NS**)
   - Other: (If "other" could you tell me what you think are the benefits(s))

Ask all respondents

3. How did you hear about tonight's meeting? (**email/text message/flier/social network/newspaper/phone call/word of mouth**)
4. Did someone contact you directly about today's meeting (**A/D/NS**)? If "Y," was the individual who contacted you **a neighbor, co-worker, family member, friend or someone else?**
5. Thinking of people who live in your community, how many of them have you talked to you about this issue?
6. Who do you normally talk to regarding this issue? **Neighbor(s), friend(s), family member(s), co-workers or someone else** (MARK all that apply)
7. On a scale of 1 to 5 (**1 = lowest to 5 = highest**), how informed are you about this issue?
8. Person's birth year.
9. Where do you live? (**TOWN/TOWNSHIP/MUNICIPALITY**)
10. Would you like to provide any additional comments?
11. Would you like us to email the results of the survey? If so, could you please provide us with an email address to reach you.

Background Reading Questions to Prep for Conducting Field Survey

Meyer, David. S. 2006. *Politics of Protest.* **New York: Oxford University Press, Chaps 2–4.**
   Meyer, Chapter 2 Why Protest

- Explain the cycles of protest, how do political opportunities fit into the cycles?
- What is the relationship between activists and institutional authorities in explaining the impact of protest movements?

- How does the concept of Eisinger's work on the structure of political systems condition levels of opportunities that stem from protest movements?
- What are the reasons to organize? How does the work of Piven and Cloward help explain variation in organization structure, capacity, and outcomes? Why do social movements form? How are organizations maintain within social movements?
- What are the major obstacles to successful vs unsuccessful movements (e.g. ideological v routinization)?

Meyer, Politics of Protest: Chapter 3: Becoming an Activist

- Why are circumstances and opportunities so important in understanding the motives as to why individuals join social movements?
- What does the author mean that in understanding the diversity of social movements?
- How does Meyer define the concept of being an "Activist?" Is his definition too broad or narrow? What would you add to this concept?
- Why are focusing on individuals "who come and go" one of the most important factors in understanding the dynamics of a social movement?
- Who are movement professionals, entrepreneurs, and anomics? What is their role and why are they important to understanding movements?
- What is the best predictor as to why an individual would engage in a political movement?
- How are social movements mobilized? Who comprises these movements?
- What are the key characteristics for finding people to participate in social movements (e.g. slack time, occupation, networks, etc.)?
- Why is acting collective so important to individual identity within social movements?
- How is a common identity and vocabulary developed among activists, organizers, and loosely affiliated members of social movements? Why are these connections important to understanding the dynamics of the movement?
- What are the challenges for activists who become professionals? How do causes and movements adapt, change, or resist according to one's ability to make a living organizing?

Meyer, Chapter 4 Individuals, Movements Organizations and Coalitions

- What does Meyer mean when he writes "that the translation of a grievance into collective action isn't automatic or unproblematic." How has scholarly literature addressed this process?
- What are the functions of social movements?
- What are the primary motives for maintaining an organization and for promoting its stability?
- What is the "iron law of oligarchy" and why is this important to understanding the organizational dynamics of social movements?
- What does Meyer mean "the form an organization takes has important effects on its prospects for survival, as well as its politics and potential influence?"

- What does Meyer mean when "[b]oth political progress and organizational maintenance required a two-track strategy?"
- Describe the process by which coalition and social movements flourish and end? What are the organizational elements that lead to success and what opportunities are needed in order for this process to work?

Gerlach, Luther. 2001. The structure of social movements: Environmental activism and its opponents. Networks and Netwars: The Future of Terror, Crime, and Militancy.

- What is the most common form of social movements presented (SPIN)
- How does he define "segmentary" elements to movements?
- Why do groups divide or splinter? "
- What are counter-movements and how do they respond to environmental movements?
- How does the author define the notion of polycentric features to movements?
- How does the Network component fit into the scheme for understanding social movements?
- How do linkages impact networks and movements?
- What are integrating factors for creating networked movements?

References

1. Collier D (1999) Data, field work, and extracting new ideas at close range. Newsl Organ Sect Comp Polit Am Polit Sci Assoc 10(1): 1–6
2. Hsueh R, Jensenius F, Newsome A (2014) Fieldwork in political science: Encountering challenges and crafting solutions: Introduction. PS: Polit Sci & Polit 47(2): 391–393. https://doi.org/10.1017/s1049096514000262
3. Hurdle J (2018) Judge grants first eminent domain case to PennEast in Pennsylvania NPR. https://stateimpact.npr.org/pennsylvania/2018/12/10/judge-grants-first-eminent-domain-case-to-penneast-in-pennsylvania/
4. Gerlach L (2001) The structure of social movements: Environmental activism and its opponents. In: Arquilla J, Ronfeldt D (eds) Networks and Netwars: The Future of Terror, Crime, and Militancy. Rand Corporation, Santa Monica
5. Johnson JB, Reynolds HT, Mycoff JD (2008) Political Science Research Methods, 6th Edition. CQ Press, Washington DC
6. Lower Saucon Township, PA (2015) PennEast project timeline. Retrieved on November 29, 2019 from http://www.lowersaucontownship.org/penneast/pepp3.pdf
7. Meyer DS (2006) Politics of protest. Oxford University Press, Oxford
8. Skrapits E (2015) Q&A with PennEast Representatives. Citizen Voice. Retrieved from https://www.citizensvoice.com/news/q-a-with-penneast-representatives-1.1977179
9. Sklavounos S, Rigas F (2006) Estimation of safety distances in the vicinity of fuel gas pipelines. J Loss Prev Process Ind 19(3): 24–31
10. US EPA (2020, March 9) Basic information on CWA Section 201 Certification US Environmental Protection Agency. Retrieved from https://www.epa.gov/cwa-401/basic-information-cwa-section-401-certification

11. Vinten G (1994) Participant observation: A model for organizational investigation? J Manag Psychol 9(2): 30–38
12. Woliver L (2002) Ethical dilemmas in personal interviewing. PS: Polit Sci & Polit 35(4): 677–678. https://doi.org/10.1017/s1049096502001154

Chapter 6
Extractive Messaging: A Critical Communicative Approach to Pipeline Pedagogy

Jessica L. Rich

Abstract The chapter examines approaches to teaching about community responses to energy development from a communication perspective. Communication about extraction and the environment is embedded in systems of power that often privilege technical and scientific knowledge over local and cultural knowledge. The chapter presents a lesson developed for an environmental communication course in which case studies from oil and gas debates in the western United States were used as the basis for student learning. Multimedia activities of meaning-making processes encourage students to examine how discourses have the capacity to legitimize technical knowledge over experiential knowledge. I argue that critical communicative approaches to pedagogical methods strengthen students capacity for analyzing messages produced by policy-makers, industry representatives, and community members. Critical thinking skills further are enhanced through an examination of systems of power and the ways that knowledge is valued differently in the environmental decision-making process. This chapter explores opportunities and challenges encountered, as well as lessons learned. Conflicts over energy development act as rich case studies for students to explore messages produced by oil and gas companies and by the communities that resist these projects.

Keywords Environmental communication · Critical communication pedagogy · Pipelines · Scientific knowledge · Local knowledge

---

J. L. Rich (✉)
Independent Scholar, Boulder, CO, USA

© Springer Nature Switzerland AG 2021
V. Banschbach and J. L. Rich (eds.), *Pipeline Pedagogy: Teaching About Energy and Environmental Justice Contestations*, AESS Interdisciplinary Environmental Studies and Sciences Series,
https://doi.org/10.1007/978-3-030-65979-0_6

6.1 Introduction

In this chapter, I present an in-depth exploration of a lesson that takes a critical communication approach to teaching about oil and gas development, examining the relationship between communities, power, and expertise in environmental decision-making processes. The lesson can be used remotely for instructors interested in incorporating pipeline pedagogy into their classrooms but whose campuses are not situated in close proximity to pipeline projects. The courses for which I developed the lesson have been taught in multiple regions and institutions of the United States, including a large university in the American West and a small liberal arts college on the East Coast. The readings and activities used for the lesson encourage students to challenge technical approaches to public participation in environmental decision-making that often favor economic and scientific knowledge and may discount cultural, local, and indigenous ways of knowing.

The primary case study used for the lesson is drawn from the struggle for community rights in opposition to oil and gas development in Boulder County, Colorado. The chapter presumes that debates related to fracking inherently are connected to pipeline discourses, as are stories of gas line and well leaks that occur in local communities due to outdated infrastructure ([9], [17]). Residents of multiple towns in Boulder County organized against drilling after the Colorado Supreme Court made invalid local bans on unconventional drilling in a May 2016 ruling [25]. As a researcher living in Colorado at the time, I attended public forums and hearings in order to follow the community responses to the Supreme Court's decision. As a resident of Boulder County, I also participated in direct action trainings organized by local community members to learn potential strategies for peaceful protesting. The lesson in this chapter was first developed for a course taught in Colorado, an ongoing hotbed of oil and gas activity.

The lesson I present in this chapter draws from my experiences living in a region with a long history of extraction, as well as my scholarly research on the community responses to unconventional drilling. Both courses in which I taught the lesson were developed on the topic of environmental communication. The activities and readings associated with the lesson emphasize the importance of considering how pipeline discourses expand beyond the development of major projects. Teaching about debates over oil and gas development necessitates discussions of local politics and geopolitics, social and cultural differences that exist within communities and between them, as well as the understanding that knowledge and expertise are not relegated to formal training or professional spheres. To develop lessons and campus efforts that address these debates, it is critical to look to diverse ways of knowing that emerge from local and embodied experiences.

My chapter begins with a review of literature focused on communicative approaches to examining environmental conflict. I draw from literature that takes a critical perspective to interrogate how the preference for technical discourses maintains uneven access to environmental decision-making processes. Next, I explore in detail the case study and related activities used in my environmental communication courses, in which students were asked to analyze a public hearing screened in the classroom. I conclude with reflections on the significance of incorporating critical communication pedagogy to teach about public participation in environmental decision-making in the classroom.

6.2 Developing Critical Communication Pipeline Pedagogy

Critical communication pedagogy (CCP) argues that identity, power, and knowledge production should be core attributes of higher-education course development [12]. I suggest that these same tenets are central to teaching about pipeline debates. Developed as a response to normative assumptions about identity and power in the communication classroom, CCP can extend beyond communication studies to environmental studies and sciences, fields that increasingly recognize the importance of intersectionality and justice in research, teaching, and practice [15]. Critical analyses of messages produced throughout the decision-making process—from project planning to inviting the public to comment to project authorization—offer opportunities for strengthening students' capacity for critically examining traditional models of environmental governance that limit knowledge produced from a diversity of standpoints.

Examining decision-making processes as communicative, *meaning-making processes* encourages students to analyze the messages produced by governing procedures used to address environmental conflicts. As established in the National Environmental Policy Act of 1970 (NEPA), the public in the United States holds the right to be informed about and to comment on decisions of environmental concern ([22, p. 295]). I turn to communication studies literature to examine how meaning-making shapes understanding of power, expertise, and ways of knowing in community responses to oil and gas development and how the formal mechanisms involved in decision-making construct spaces that privilege technical expertise while dismissing community-based, experiential knowledge. Analyses of these processes develop students' critical thinking skills by challenging decision-making as neutral, objective, and unbiased.

Communication research examines processes of meaning-making and how social interaction shapes the human experience of the world. Communication studies informed by critical-cultural research emphasize the role of power and how systems of privilege are reproduced through symbolic interaction [18] and the role of ideology as produced not by individual behaviors, "but in the interplay of symbols and meanings that make up the lived-world of the individual" ([20, p. 295]). Teaching communication from a critical perspective presents opportunities for students to analyze how power operates through the dialogue and discourses produced in the environmental decision-making process. US-based environmental decision-making requires government oversight, leaving communities to depend on formal authority at various levels to authorize public participation in policies impacting the public good.

As a sub-field within the discipline of communication, environmental communication can be defined as an academic and professional practice. Practices, such as environmental policy analysis and advocacy, are examples of professional activities in the field. Environmental communication scholars investigate the discourses that give social, political, scientific, and historical meaning to relationships between the human and the natural world. Communication about nature "powerfully [affects] our perceptions of the living world; in turn, these perceptions help shape how we define our relations with and within nature and how we act toward nature" ([19, pp. 345–349]). Contestations concerning contemporary environmental issues, such as oil and gas development and pipeline projects, provide significant source material for communication scholars interested in how official decision-making processes come to define and legitimize human–nature relationships in their various forms.

Given the urgency of contemporary environmental problems, such as human-induced climate change, environmental communication offers researchers and teachers with valuable tools for challenging environmental policies and industry practices that maintain a dangerous status quo. Cox [5] defines environmental communication as a "crisis discipline" with the responsibility to demand transparent decision-making processes that invite equitable public participation. Cox further suggests that the discipline holds an ethical duty to interrogate policies that are harmful to environmental and public health (pp. 15–17).

Environmental decision-making often relies on expertise informed by technical and scientific knowledge as more objective than other forms, such as produced knowledge by members of frontline communities impacted by the decisions themselves ([14], [21]). Decision-makers frame assessments from scientists, economists, and government officials as "unbiased and accurate technical advice" ([23, p. 106]) compared with public input, deemed as "irrational, subjective, ignorant fears" ([21, p. 90]). Fischer [10] defines local knowledge as:

> knowledge about a local context or setting, including empirical knowledge of specific characteristics, circumstances, events, and relationships, as well as the normative understandings of their meaning…a type of knowledge that owes its status not to distinctive professional methods but to casual empiricism, thoughtful reflection, and common sense (p. 146).

While scientific data is vital to studies of environmental problems and their impacts, decisions that privilege such information over the knowledge grounded in the experiences of members of the public risk the possibility of erasing meaningful, community-based contributions. Vital to maintaining a healthy public sphere is the inclusion of democratic deliberation among its participants ([10], [14]). Official processes that discount the value of local knowledge diminish the possibilities for productive communication between publics and their representative government.

Knowledge becomes legitimated through decision-making processes. The preference for technical language, that which is used in scientific or legal fields, perpetuates the uneven distribution of power in environmental decision-making. Peterson [21] defines two forms of discourses, technological and creative, used in the participation of policy decisions. Technological discourses favor quantitative data as the primary support for rational arguments and "exclude evidence and arguments drawn from nonnumerically defined experience" ([21, p. 101]). Creative discourses, on the other hand, invite arguments that stem from experiential knowledge and diverse ways of knowing. Peterson notes that technological discourse "predetermines" which information will be accepted as evidence, while creative discourses "[encourage] participants to pursue previously unforeseen alternatives" (p. 101). Individuals whose training falls within the parameters of a given form of expertise, typically those with the authority to oversee the governing operations of decision-making processes, ultimately define the parameters of expertise as well as which knowledge is or is not legitimate.

Technical experts dominate decision-making in energy policy [14]. Pipeline projects, for example, submit data compiled by engineers, geologists, attorneys, and economists, as evidence of safe practices, legality, and benefits to the financial well-being of communities. Technological discourses form the foundations of information to which decision-makers turn when developing pipeline policy, while the experiential and embodied knowledge of local communities is secondary at best, contributed later in the process during public comment periods. Known as a decide-announce-defend (D.A.D.) model, policy-makers release a decision on a project and then justify their plans if the public responds with concerns [6]. Although procedures are in place to document public comments during specified periods throughout the environmental decision-making process, oil and gas projects often move forward despite public outcry.

Rather than imagining expertise as occurring from the top down—for example, from scientist to laypeople—a communicative approach grounded in a critical pedagogy interrogates such a model. Teaching from this approach suggests that traditional modes of decision-making sustain the privilege of professional expertise and marginalize the knowledge of individuals and groups who base their expertise on experiences of the body and place (i.e., environmental stressors such as illness, mental and emotional burdens of living near polluted sites). "Critical communication pedagogy is…about dialogue or engagement between various constituencies, dialogue that builds spaces for transforming the world as it is in favor of a collaborative vision of what could be" ([8, p. 6]). Processes that integrate science with the local knowledge and creative discourses of affected communities: 1. recognize the significance of democratic participation in environmental decisions and challenge the alienation and marginalization of individuals and groups, 2. legitimate decisions in the eyes of the public, and 3. make technical expertise and data more relevant to a diversity of stakeholders [10]. Decision-making that is informed by both technical and experiential knowledge moves governance toward more socio-ecological just futures.

6.3 Analyzing Extractive Messages

The lesson in this section was taught in two sections of the course, "Introduction to Environmental Communication." The first iteration took place in Spring 2017 in a large, research university in the American West. The second was taught in Fall 2017 at a small, liberal arts college in New England. Presenting discourses from pipeline cases allowed students to examine the cultural messages associated with environmental movements and decision-making processes. Examining narratives produced by public hearings over oil and gas development encourage students to think critically about systems of power. Throughout the lessons and discussions about pipeline controversies, students examine the cultural messages produced in pipeline debates and analyzed the discourse of "expertise" that so often privileges technical knowledge over the local, embodied knowledge of communities substantially impacted by oil and gas development. Throughout the lesson, students challenge conventional processes of governance to imagine the possibilities and decision outcomes if experiential knowledge were given equal standing to technical expertise in the presented case. The activities involved in the lesson ask students, some of whom may not be accustomed to questioning governmental processes, to grapple with the tension and discomfort of critiquing the formal procedures of local, state, and federal authority.

At the start of the lesson, multiple readings familiarize students with concepts related to environmental conflict, environmental justice, and diverse forms of expertise. The primary textbook used for the courses involved in this lesson is *Environmental Communication and the Public Sphere* [22], which provides extensive, accessible material on the history, strength and weaknesses of environmental decision-making processes in the United States. Supplemental readings include, "Environmentalism and Social Justice" [2], which acts as a primer on environmental justice and the disproportionate impact of polluting industries forced onto communities of color; "Sharing the Earth: The Rhetoric of Sustainable Development" [21], in which the author offers a rich analysis of the technical procedures that discount indigenous and experiential knowledge in a case study of buffalo management in Canada's Wood Buffalo National Park; and "Citizens as Local Experts" [10], which argues for the importance of integrating public input as a form of expertise in environmental decision-making. The readings build a foundation from which to analyze cultural messages and a shared language with which to participate in classroom discussions.

6.3.1 Case Study Background

The case study used for the lesson is based on the context of debates over oil and gas development in Boulder County, Colorado. A series of public hearings followed Colorado Supreme Court's vote to invalidate moratoria on fracking in the towns of Longmont and Fort Collins in 2016 [25]. Based on the principle of pre-emption, the Court argued that local decisions to ban fracking interfere with state laws on drilling. The decision opened Boulder County, the focus of this case study, and other counties in the state to oil and gas development. Community groups, some of which had been organized in years prior, strengthened their opposition and became more public in their efforts to prevent drilling near residences, schools, and businesses. Extractive industries have long operated in the State of Colorado, but Boulder County held oil and gas at bay as counties to the east saw thousands of wells permitted to operate. Oil and gas development necessitates the construction of small pipelines, wells, and drill pads in communities and the disruption of everyday life for community members.

Boulder County commissioners held a series of public hearings after the Supreme Court decision, inviting residents to submit written and oral comments. A short-term moratorium was established to expire in May 2017, giving the County time to develop regulations, which were reviewed at an open meeting in November 2016. County staff associated with public health and land use researched and offered 25 amendments on the regulations, including requiring baseline testing on water supplies within a half-mile of a drill site, periodic testing of water sources one mile from a drill site, and requiring a distance of 150 feet between buildings and pipelines. A Commission meeting on March 14, 2017 saw a public review of the amendments, as well as the opportunity for public comment, excerpts from which form the basis for students' analysis in the lesson.

6.3.2 Examining Power in Energy Discourses

Boulder County is a useful case for teaching about community responses to energy development and public participation. The community group, East Boulder County United, was formed by Lafayette, Colorado residents and was one of the more visible and organized throughout the public comment period, appearing at the formal hearings and participating in protests. Still active, the group was essential in advocating for the Climate Bill of Rights, which declares among its protections for the people of the City of Lafayette, the right to a healthy climate, the right to governance by the people, and the people's right to defend against risks posed by the extraction of coal, oil, and gas [4]. Some participants in the group had been involved in the Standing Rock protests and language in the Bill reflects similar values. In addition to scholarly articles related to environmental decision-making, expertise, and environmental justice, students are assigned to read the local ordinance introduced by the City of Lafayette in 2017.

The classes also viewed excerpts from the March 14, 2017 public hearing on oil and gas development in Boulder County, Colorado ([1], [24]). Students analyzed how meanings are produced within a public hearing, paying particular attention to different representations of expertise, effects of space on social relations, as well as systems of value that emerge in the hearing. The lesson took place over two 75-minute class meetings in one instance and one three-hour class meeting in the other instance. The hearing remains publicly available through the Boulder County website. Due to the length of the video at almost five hours, I chose excerpts from the hearing to capture the procedural format of the hearing and a selection of different views offered by the public speakers. Portions of the hearing shown included introductory remarks by leaders of the hearing, presentations by County staff, comments made by members of the public (the majority of whom opposed drilling), as well as representatives from the oil and gas industry, who spoke in favor of development.

The rules and procedures of public hearings introduce a significant opportunity for examining varying levels of power among participants involved in decision-making. Students viewed the opening comments by the commissioners at the hearing. The individuals presiding over the hearing, Commissioners Deb Gardner, Elise Jones, and Cindy Domenico, reviewed the hearing agenda and established the rules of the proceedings. A three-minute time limit was given to public speakers, and the class discussed the challenge of developing a persuasive argument within such a short time-frame. Students commented on the formal authority that commissioners held over the audience's behavior. Important insights that students made included the capacity of commissioners to oversee "civil conduct," preventing audience members from voicing opinions outside of their allotted time for comments. For example, the audience was not allowed to "cheer, boo, hiss, cry out loud or applaud" during testimonies, nor were signs permitted in the hearing room. Such rules are commonplace in formal settings, communicating the formal ownership of power within the space and also defining which practices are appropriate or inappropriate among participants. In addition, the Commissioners point out how some members of the public voiced disappointment that the hearing was held during the day, preventing those who work during business hours from participating. The Commission responded to the criticism by running the public comment period into the evening; however, one student noted that the Commissioners apologized for not making the scheduling more clear, lacking sensitivity to a diversity of work schedules that do not fit within the confines of the nine-to-five workday. The rules that govern decision-making processes are useful for examining how dissenting voices become constituted as out of place within the formal setting of the hearing room.

6.3.3 *Exploring the Production of Expertise*

Technical expertise was central to the opening of the testimonies in the case. An extended period was given to the Boulder County staff for a formal presentation before the public comments began. Staff from the County's Land Use Department reviewed proposed regulations on oil and gas developed to protect public health. An air quality and public health coordinator presented techniques for monitoring air quality and standards required by the Environmental Protection Agency. An attorney for the county also spoke on the different levels of authority that cities and counties hold to govern land use policies and for developing future moratoria on oil and gas development. The screening of the hearing allowed for the class to pause the video and hold a brief discussion between different points in the hearing. After viewing the formal presentations, we discussed how a priority was placed on environmental and legal expertise, with staff allotted time to present their evaluations before leading into the public testimonies.

Next, we viewed comments made by several Boulder County residents, and students were given time to respond to the differences in presentation of technical expertise and local expertise. For example, a student enthusiastically pointed out how the first speaker began her comments by discussing a common indigenous custom to "acknowledge our connection to the natural world" at formal gatherings. The speaker invited those gathered to imagine themselves near the local Boulder Creek and, quoting from an Iroquois greeting, to "give thanks to the clear water, the living soil, the warmth of the sun, the fresh air we breathe, the insects and animals, the giving trees and plants, and the great life force." The language stands in stark contrast to the review of environmental, economic, legal, and health data presented by Boulder County staff at the start of the hearing. The class compared the perception of nature as object, as defined through terms such as natural resources and land use, with the Iroquois greeting, which places the more-than-human community on equal footing with the human community. In addition to analyzing expertise in the hearing, students examined the values that emerged from speaker comments and whether the arguments were persuasive. Questions that students were asked to consider included: What are examples of technical expertise that you notice? How does technical knowledge operate in the comments? What are examples of local knowledge that you notice? How does experiential knowledge operate? The class applied concepts, such as technical and experiential expertise, to the discourses produced by the hearing, to examine how the discourses expose value systems such as those that emerge from Western and indigenous worldviews.

Many of the proceeding speakers spoke from personal standpoints, voicing concerns for their homes and children's schools if oil and gas development were to be extended into Boulder County. One particular moment that drew student interest was a resident from Lafayette who represented the organization, East Boulder County United, and identified herself as a Boulder County Protector, language mirroring that of participants in the Standing Rock protests who used the label Water Protectors to define their commitment to insulating the Missouri and Mississippi Rivers from the harm promised by the Dakota Access Pipeline [13]. In her comments, the speaker defines Boulder County Protectors as being trained in direct action to protect residents from fracking and climate change, which the organization sees as inexplicably linked. She expresses feeling failed by local government to protect their residents, and during her testimony, uses her time to lead the audience in chanting, "For our Earth, we will stand, hold the power in our hands. Drills and pipelines, we will fight. Healthy climate is our right." She ends with the question, "Commissioners, are you going to poison us, or protect us?" Audience applause follow the speaker's comments, which a Commissioner quickly halts.

Many students shifted in their seats and some giggled during the Boulder County Protector's time at the podium. When pressed to reflect on how they felt about the comments, some said the presentation seemed "odd" and "embarrassing," because the speaker seemed out of place in this formal space. One student spoke more at length to say that while the chanting made him feel awkward for the speaker, he also revealed that the chanting gave him "chills." He shared that he was impressed with how brave she was for speaking out in the room full of people. We discussed the practices involved in constructing community members as "out of place," including the formal rules that limit forms of expression that may challenge procedures set by the ruling authority. The occupation of this formal space by members of public worked to challenge the formal authority that typically prevents such meaningful forms of resistance.

Comparing the verbal as well as nonverbal communication practices of local residents, County staff, Commissioners, and industry representatives provided another opportunity for students to critically analyze the processes that constitute technical knowledge as rational and invalidates experiential knowledge. Students were interested in commenting on the nonverbal messages that different styles of dress communicated to viewers. For example, local residents tended to wear clothing perceived as more casual in comparison to the business attire worn by those representing the County or industry. A staff member from the Colorado Attorney General's Office who also represented the Colorado Oil and Gas Conservation Commission, a middle-aged, white man, wore a dark suit and tie and used his time to challenge whether proposed rules on development in the County are preempted by State law. During his opening remarks, audience members can be heard stirring, and a Commissioner scolds, "Please keep your comments to yourself." As a class, we discussed the "fit" of the industry representative in the hearing room and how his language and professional dress appeared to be legitimated by the proceedings and the difficulty that local residents faced when arguing their case against industry operations, no matter how dangerous, that are supported by law. The lawyer's presentation marks a moment in which to illustrate the differences in the language of technical knowledge and experiential knowledge, as well as additional factors that maintain the power of technical expertise.

Screening the hearings also provides an opportunity to investigate how space, another form of nonverbal communication, enables and constrains social relations and gendered, raced, and classed identities [16]. For example, students were asked to examine the setting of the room in which the hearing took place and how the space communicates power to participants. As is typical in public meetings such as this, the commissioners sat at the front of the room with members of the public sitting in chairs and standing one-by-one at a podium to present their comments. Students noted the formidable wooden dais behind which the commissioners were positioned and the mass-produced chairs on the hearing floor where the audience were seated.

In some instances, members of the public have occupied this space differently, disrupting the formality of the proceedings. I showed students photographs of protests that took place in the same room. For example, one photo showed members of a community rights group carrying out a "die-in" at a different commission hearing in Boulder County [7]. Individuals dressed in hazmat suits while others carry tombstones declaring, "Fracking Kills All." In contentious cases, members of the public can feel especially frustrated by the decision-making process, especially when they feel that their voices aren't represented or that their representatives aren't listening or acting in their best interest.

The in-class analysis of public hearings explores how expertise can take different forms but also how experts are valued differently in the decision-making process. "Along with the presumption of authority comes a belief in professional hegemony, 'the belief that in the area of technical expertise, which is defined by the profession, issues are to be decided by rigorous standards of evidence and argument, which again the profession defines'" ([23, p. 107]). Formal authorities create rules for decorum, oversee procedures, and ultimately legitimize expertise. Teaching from public hearings can reveal the disciplinary power, borrowing a concept from Foucault [11], of local and state bureaucracies to privilege technocratic knowledge over the local. The processes of constituting disciplinary power become evident in the critical examination of discourses and spaces of the hearing room.

As mentioned earlier, the lesson was taught in two regions of the United States, the West and New England. The regional distinctions posed somewhat of a challenge when screening the hearing in the two classes. I found that students connected differently with pipeline controversies, particularly when it came to environmental politics. Students in the West were familiar with local debates over oil and gas development. Most grew up in the region and many viewed oil and gas as embedded in the West's identity and vital to regional economies. In New England, students tended to perceive pipelines and fracking as a more distant threat, although one student from New Hampshire stated how the Boulder case reminded her of the termination of the Access Northeast pipeline project in New Hampshire in 2017 [3]. Despite the differences between the two regions, classroom discussions revealed that students felt strongly that oil and gas development was not worth the risks to environmental and public health or at least that it should be regulated more than it is.

In addition, with the Boulder County example as a guiding case study, students from the two regions developed critical analytic skills that crossed political standpoints. Students learned how environmental decision-making is never neutral. Rather, the process sustains power and value systems with the effect of discounting the experiences of local communities. Further, the decision-making processes reinforce problematic narratives and discourage dialogue among participants. Debates rage in part due to the construction of a false binary built between protagonists and antagonists, proponents of industry and their opponents. As Fischer [10] notes, environmental decision-making flourishes if the process recognizes the value that experiential knowledge can offer when taken into account with technical knowledge. Studying decision-making as a process of meaning-making and knowledge production can expose how one form of expertise becomes valued over another. The analysis also opens students to a consideration of how to pursue more equitable and just systems of governance.

6.4 Conclusion

Pipeline cases offer environmental studies classrooms an opportunity to investigate questions about human relationships with the earth, energy, and one another. Communication approaches are important in the teaching of pipeline conflicts because they assist instructors in incorporating pipeline pedagogy into classrooms, despite geographical proximity to development. A communicative approach allows instructors to discuss how humans shape understanding of nature through language and how this understanding has led to destructive relationships with the environment and unequal power relations in decision-making processes.

Discounting experiential knowledge limits the contributions of community members who live on the frontline of oil and gas development. Further, communities of color are disproportionately at risk of environmental stressors, meaning these communities often are key, expert voices in discussions of air quality, water contamination, and public health. The historical destruction of environmental justice communities by polluting industries reinforces the responsibility of groups in positions of power and privilege to listen to those on the frontlines. Class discussions used to critique the dominant ideologies that support a limited view on what constitutes expertise—and who constitutes an expert—encourage students to consider how different ways of knowing are vital to a deliberative democracy.

References

1. Boulder County (2017, March 14) Amendments to oil and gas development regulations Retrieved from http://bouldercountyco.iqm2.com/Citizens/default.aspx
2. Bullard RD (1994) Dumping in Dixie: Race, class, and environmental quality. Westview Press, Boulder CO
3. Chesto J (2017, June 29) Lacking financing, utilities put $3 billion natural gas pipeline plan on hold. The Boston Globe. Retrieved from https://www.bostonglobe.com/business/2017/06/29/utilities-withdraw-plan-for-billion-natural-gas-pipeline-expansion/oO7zbTYmUIMWVmpjNItQ3H/story.html
4. City Council of the City of Lafayette (2017, March 21) An Ordinance of the City Council of the City of Lafayette, Colorado, Enacting the Climate Bill of Rights and Protections. Ordinance No. 02, Series 2017
5. Cox R (2007) Nature's "crisis discipline:" Does environmental communication have an ethical duty? Environmental Communication 1(1):5–20
6. Endres D (2012) Sacred land or national sacrifice zone: The role of values in the Yucca Mountain participation process. Environmental Communication 6(3):328–345
7. East Boulder County United (2017, October 11) Press statement on die-in at the Boulder County Commissioners. https://www.eastbocounited.org/single-post/2017/10/03/Press-Statement-on-Die-In-at-the-Boulder-County-Commissioners
8. Fassett DL, Warren JT (2008) Pedagogy of relevance: A critical communication pedagogy agenda for the 'basic' course. Basic Communication Course Annual 20(6):1–34
9. Finley B (2017, May 2) Deadly Firestone explosion caused by odorless gas leaking from cut gas flow pipeline. The Denver Post. Retrieved from https://www.denverpost.com/2017/05/02/firestone-explosion-cause-cut-gas-line/
10. Fischer F (2000) Citizens, experts, and the environment: The politics of local knowledge. Duke University Press, Durham
11. Foucault M (1995) Discipline and punish. (Trans: Sheridan A). Vintage Books, New York
12. Golsan KB, Rudick CK (2018) Critical communication pedagogy in/about/through the communication classroom. J Communication Pedagogy 1(1):16–19
13. Kennedy M (2016, September 9) Judge rules that construction can proceed on Dakota Access Pipeline. National Public Radio. Retrieved from https://www.npr.org/sections/thetwo-way/2016/09/09/493280504/judge-rules-that-construction-can-proceed-on-dakota-access-pipeline
14. Kinsella WJ (2004) Public expertise: A foundation for citizen participation in energy and environmental decisions. In: Depoe SP, Elsenbeer MA (eds) Communication and public participation in environmental decision making. SUNY Press, Albany, pp 83–95

15. Lloro-Bidart T, Finewood MH (2018) Intersectional feminism for the environmental studies and sciences: Looking inward and outward. J Environmental Studies and Sciences 8(2):142–151
16. Massey D (1994) Space, place, and gender. University of Minnesota Press, Minneapolis
17. Mauro M (2019, October 14) Abandoned oil and gas well in Broomfield leaking methane. KDVR. Retrieved from https://kdvr.com/2019/10/14/abandoned-oil-and-gas-well-in-broomfield-leaking-methane/
18. Mead GH, Morris CW, Huebner DR, Joas H (2015) Mind, self, and society: The, definitive edn. The University of Chicago Press, Chicago
19. Milstein T (2009) Environmental communication theories. In: Littlejohn SW, Foss KA (eds) Encyclopedia of Communication Theory. Sage, Thousand Oaks, pp 345–349
20. Mumby D (1989) Ideology & the social construction of meaning: A communication perspective. Communication Quarterly 37(4):291–304
21. Peterson TR (1997) Sharing the Earth: The rhetoric of sustainable development. University of South Carolina Press, Columbia
22. Pezzullo PC, Cox RJ (2019) Environmental communication and the public sphere, 5th edn. Sage, Thousand Oaks
23. Richardson M, Sherman J, Gismondi M (1993) Winning back the words: Confronting experts in an environmental public hearing. University of Toronto Press, Toronto
24. The Nation Report (2017, March 14) Boulder County residents tell commissioners they will never accept fracking. Retrieved from http://www.thenationreport.org/boulder-county-residents-tell-commissioners-they-will-never-accept-fracking/
25. Wines M (2016, May 2) Colorado Court strikes down local bans on fracking. The New York Times. Retrieved from https://www.nytimes.com/2016/05/03/us/colorado-court-strikes-down-local-bans-on-fracking.html

Part III
Mobilizing Pipeline Politics

Chapter 7
Mountain Valley Pipeline: A Case Study in Local Resistance and Mobilization

Diana Christopulos

Abstract This case study shows how landowners, environmentalists, and others across the political spectrum are creating major challenges for Mountain Valley Pipeline, a 303-mile long interstate fracked natural gas pipeline project in Virginia and West Virginia. College faculty, students, graduates, and retirees are playing major roles in this effort. Highlights are reviewed here, along with nine principles for opposing pipeline projects, told from the viewpoint of an active participant. At the time of this writing (May 2020), the project is at least two years behind schedule and almost 70% above budget. MVP lost two sets of federal permits in 2018 and another in 2019 and has not yet recovered them. Opponents are hopeful that these delays will allow federal laws and declining international markets for liquid natural gas (LNG) to catch up with an industry that ignores both private property rights and environmental protection.

Keywords Pipeline protest · Mountain Valley Pipeline · Appalachia · Coalition movements · Property rights

7.1 Introduction and Principles of Protest

Fighting pipelines is an ultramarathon punctuated by dozens of sprints. If completed, Mountain Valley Pipeline (MVP)[1] would be a 303-mile long, 42-inch underground

---

[1] The name of this pipeline is a little inside joke by the Pittsburgh-based EQT fracking company that began the project. The letters MVP could also stand for Most Valuable Player, and all of the compressor stations in the project are named after former Pittsburgh Steelers who won that honor in the National Football League. At a June 2017 meeting organized by the National Park Service in Salem, Virginia to discuss MVP and the Appalachian Trail with relevant agencies and intervenors, the MVP representative urged everyone to make friends during lunch and discuss something all could agree on, "like the Steelers." He had no sense that the Roanoke region is devoted to Virginia Tech, not teams in Pittsburgh.

D. Christopulos (✉)
Salem, VA 24153, USA

pipeline in West Virginia and Virginia full of highly explosive natural gas under approximately 1,440 lb per square inch of pressure. It was proposed in 2014, received its federal and state permits in 2017 and began construction on a very hazardous route in 2018. The project has been severely hampered by landslides, erosion and the loss of three sets of federal permits related to water pollution, endangered species and impacts on the Appalachian Trail. The project is over two years behind schedule, and costs have escalated more than 70%. EQT Corporation (EQT), the Pittsburgh, Pennsylvania fracking company that started MVP to get its stranded oversupply of natural gas to export markets, is selling its share in the project, and its bonds received a junk rating from Moody's in early 2020 [126].

This case study focuses primarily on the 100 miles of the MVP route in Virginia, especially in areas near the Roanoke and New rivers. Health care, education, and tourism are the region's major economic drivers. The Roanoke Valley has a population of more than 300,000 people, with three major hospitals and 12 colleges. The nearby New River Valley includes Virginia Tech, a public university with over 34,000 students and almost 2,000 faculty.

While I do not live on the MVP route, I can see it from my neighborhood. Much of my involvement has been around impacts to the Appalachian Trail. In my professional life, I have been a college professor, a nonprofit executive and the owner of an international management consulting business. I came of age during the height of civil rights and Vietnam War protests, participating in three major marches on Washington. I am the only person I know who sold Girl Scout cookies at the Pentagon in the 1950s and marched on it in the 1960s. I have always loved being outdoors and have rafted, canoed, or kayaked thousands of miles of American rivers and hiked or backpacked thousands more, including the entire Appalachian Trail. In Texas, I was the Sierra Club's volunteer State River Protection Chair in the 1990s and in Virginia I co-founded the all-volunteer Roanoke Valley Cool Cities Coalition in 2006, with a focus on climate change. I returned to the classroom in Spring 2020, co-teaching a seminar on environmental history and public policy at Hollins University.

When I mention "we" below, I am talking about a loose coalition that involves many individuals and organizations. We are not all working on everything together all the time, although many of us are in regular communication. We know and trust one another after more than five years of work. The most important local group has been Protect Our Water, Heritage, Rights (POWHR), a coalition of geographically based organizations opposed to MVP. They are informally connected to many other groups and individuals, and I count myself among that number. Most of my own work has been focused on the Appalachian Trail, Jefferson National Forest and issues that affect the cities of Salem and Roanoke.

Landowners on or near a pipeline route are its most steadfast opponents, just as victims of heart disease or cancer are the strongest proponents of prevention and cures.

I inductively developed an informal set of principles while opposing MVP. You might think of them as Saul Alinsky meets Ruth Bader Ginsberg. Alinsky led many successful efforts to organize and empower poor communities from the 1930s to the 1970s and wrote a set of 13 rules for radicals such as, "A good tactic is one your people enjoy" and "The price of a successful attack is a constructive alternative" [19]. Ginsberg is the Supreme Court Justice who graduated first in her class at Columbia Law School and wrote landmark decisions on equal rights for women and the handicapped as well as legal standing for environmental groups.

Opposing a project that costs billions of dollars and has the blessings of federal agencies, state governors, U.S. Senators and other people with power is not easy, yet our loose coalition has already accomplished a great deal with very minimal financial resources. College and university faculty, staff, retirees, and students have been active participants in opposition to MVP since it was first announced. Here is a brief overview of some principles I think we subconsciously followed, along with examples of our actions:

- **Show up.** We have attended countless public hearings and court trials at the federal, state, and local levels, providing expert testimony to officials and emotional support to landowners and other protesters.
- **Create a strong core and continually find new allies.** A core base of landowners and environmentalists has expanded to include local governments, health professionals, religious leaders, college students, state and federal elected officials, defenders of the Appalachian Trail and other pipeline opponents across the nation.
- **Learn the facts and tell the truth.** We have written or funded expert reports on the geologic dangers and negative economic, visual, cultural, and environmental impacts of MVP. While MVP frequently exaggerates its benefits and ignores negative impacts, we try to be as scientific and fact-based as possible.
- **Talk to elected officials.** We have personally met with county supervisors; city councils; state governors, delegates, and senators; as well U.S. Senators and members of Congress. These meetings have often resulted in formal statements of opposition or specific complaints against MVP, the Federal Energy Regulatory Commission (FERC) or the Virginia Department of Environmental Quality (DEQ).
- **Inform, involve, and become the media.** On the ground, observers have frequently provided tips that bring print and broadcast media coverage, including national and international coverage. Regional print and broadcast media have covered the MVP story hundreds of times and continue to do so. We have been able to publish many op ed commentaries and author news articles for emerging statewide digital newspapers such as the *Virginia Mercury* and *Blue Virginia*. A local videographer, Marino Colmano, created a series of powerful videos that helped move MVP away from a community with no escape route in case of disaster and eventually resulted in an award-winning full-length film, "Pipeline Fighters."
- **Find creative ways to protest.** Both colorful direct action and expert-driven alternative meetings have gained widespread media coverage and sympathy. Tree sitters have occupied many parts of the MVP route since early 2018, with a mother and daughter (Red and Minor Terry of Roanoke County) attracting international attention for defending their land. When MVP or the FERC held public meetings or hearings in the Roanoke region, we held competing news conferences in nearby

rooms along with clever signage and expert speakers, usually garnering more media attention than the formal events.
- **Use the legal system.** With help from national and regional environmental groups such as the Sierra Club and Appalachian Mountain Advocates, we challenged MVP in state and federal courts, successfully forcing withdrawal of the Biological Opinion rendered by the U.S. Fish and Wildlife Service as well permits issued by the U.S. Forest Service, the Bureau of Land Management, the U.S. Army Corps of Engineers. These actions have already delayed project completion by at least two years. Landowners have also challenged the constitutionality of the project in state and federal courts, so far without success.
- **Be watchdogs on the pipeline route.** To keep informed eyes on the MVP route, local residents received training as Mountain Valley Watch (MVW) to document and report MVP erosion and sedimentation violations to the Virginia DEQ and others, which the state's Attorney General used in a successful lawsuit against MVP [140].
- **Stay the course.** Fighting pipelines is exhausting. Many of us have been doing it since 2014. MVP has billions of dollars, support from federal politicians of both parties and current legal precedents that support seizure of private property to benefit private corporations. We have the truth, resolve, and commitment fed by the ruthless treatment of our land, water, citizens, and other living things. The best thing about MVP is the community of pipeline fighters it has created. We take breaks. But we come back. These projects are so wrong on so many levels that we believe legislative, judicial, and executive actions will eventually outlaw them in their current form. It will not happen unless we stay the course.

In the following sections, I present background on the MVP pipeline, including the economic landscape, permitting, and an overview of project supporters and opponents. Then, I provide a chronology of the project along with detailed descriptions of the nine principles of pipeline campaigning.

7.2 Basic MVP Facts: "Gas Molecules Flow All Over the Place"

The MVP project was started by EQT, a natural gas fracking company based in Pittsburgh. A remarkably similar project, the Atlantic Coast Pipeline (ACP), was begun at the same time Fig. 7.1. Both are responses to a glut of fracked natural gas in the Marcellus and Utica shale whose producers seek external markets. These huge, 42-inch pipelines are more than twice the size of pipelines that are already adequately supplying Virginia and the Carolinas.

MVP selected a very hazardous and destructive route, almost certainly because they thought its rural residents would be unable to mount much opposition. Of the 303 miles, 82% was forested, 74% has high landslide potential, 40% is on exceptionally steep grades (15–80%) and over 50 miles is karst, full of sinkholes, caves and

Fig. 7.1 MVP and ACP (Appalachian Mountain Advocates)

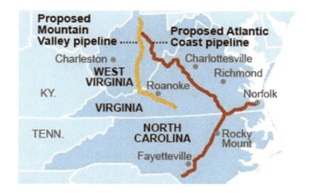

underground waterways. The entire Virginia route would be within the impact area of the very active Giles County Seismic Zone and/or the Central Virginia Seismic Zone.

7.2.1 Economics and Public Need

The primary beneficiaries of interstate natural gas pipelines are the companies that own and operate them. The FERC grants the companies the federal power of eminent domain to take and use easements from unwilling owners before paying them a nickel. The FERC also allows them to earn over 14% *net profit annually* [123]. No actual public need is required. The FERC allows developers to prove need simply by signing up subsidiaries of its own companies as long-term "customers." The 2016 "customers" and owners of MVP were all parts of the same companies Table 7.1.

Only Roanoke Gas and Consolidated Edison are public utilities. If the pipeline is completed, their local customers will foot the bill for MVP to assure profits in excess of 14% annually. Experience with similar pipelines shows that domestic customers

Table 7.1 Owners and customers of the Mountain Valley Pipeline

Owner	Customer
EQT Midstream (45.5%)	EQT (64.5%)
NextEra Energy (31%)	USG Properties Marcellus Holdings (12.5%)
Consolidated Edison (12.5%)	Consolidated Edison (12.5%)
WGL Holdings (7%)	WGL Midstream (10%)
Vega Energy Partners (3%)	
RGC Resources (1%)[a]	Roanoke Gas (0.5%)

[a]RGC Resources was not an original partner in MVP; it was added to give the appearance of meeting a local need
Reproduced from Kunkel and Sanzillo [119]

can expect to pay much higher prices [52]. The other 87% of the gas could be shipped overseas, and partner WGL has a contract to sell natural gas in India. The first time MVP came to town, a member of the Roanoke County Board of asked whether the gas in MVP was intended for export. A company representative gave an astounding reply: "Gas molecules flow all over the place" [2, 12]. Landowners and pipeline workers all report that the actual steel pipe being used for MVP was constructed in India. MVP is not a job source in Virginia. The counties it would cross generally have low unemployment rates and economies based on health care, education, and tourism rather than on manufacturing or extraction. People who live in the region are not out-of-work coal miners, and they have a history of successfully defending their mountains and valleys against power lines, industrial wind turbines, uranium mining, and other encroachments. MVP itself exists only on paper. It has no employees and gives the EQT building in Pittsburgh as its address.[2] By 2016, before MVP received any of its permits, numerous studies showed that the project was unnecessary and would likely result in high-cost natural gas, minimal economic benefits, and high economic costs to communities along the pipeline route.

7.2.2 Permitting Process

MVP needs many permits, but four appear to be most critical: two from the FERC and one from each of the state agencies responsible for enforcing the federal Clean Water Act. In Virginia this is the all-volunteer State Water Control Board (SWCB). The battles for these permits began in 2015 and appeared to end in victory for MVP in 2017.

As an interstate pipeline, MVP needs two key permits from the FERC:

- A Final Environmental Impact Statement (FEIS). MVP must comply with the National Environmental Policy Act (NEPA), a law that has some teeth (Environmental Science).
- A Certificate of Public Convenience and Necessity. This requires a vote of the presidentially appointed FERC Commissioners. The FERC has never rejected such a request if the applicant can show that they have "customers."

As FERC began writing the Draft Environmental Impact Statement (DEIS) in 2016, other state and federal agencies had to decide whether to become "cooperating agencies," which means they could use FERC's EIS instead of doing their own. Four of the agencies that decided to cooperate using the FERC EIS were:

---

[2]In the first half of 2020, MVP changed their address to 2200 Energy Drive, Canonsburg, PA 15317. This is the address of ETRN, the second EQT spinoff to assume management of the project. EQM was the other.

- The U.S. Forest Service (FS), which has to issue permits for crossing the Appalachian Trail (AT) and Jefferson National Forest (JNF).[3]
- The Bureau of Land Management (BLM), which has to agree to these FS permits.
- The U.S. Fish and Wildlife Service (F&W), which has to assure protection of threatened and endangered species.
- The U.S. Army Corps of Engineers (COE), which has to allow crossings of waterways and wetlands.

All four agencies would later have their permits withdrawn by the U.S. Fourth Circuit Court of Appeals due to serious flaws in the FEIS.

The National Park Service (NPS) was not a cooperating agency, and the FERC generally closed them out of the decision-making process. MVP would cross both the AT and the Blue Ridge Parkway (BRP), and both are NPS units. By federal law, the superintendent of BRP has the legal authority to issue a permit for a crossing by an interstate natural gas pipeline but the superintendent of the AT does not, unless the crossing is on private or state property. On some parts of the trail in national forests, the FS may have the authority to allow an AT crossing, a question the U.S. Supreme Court is reviewing at this writing. MVP's strategy was to work closely with the FS and the BRP while shutting the NPS and its nonprofit management partners out of the discussion.

In addition to federal permits, MVP needs permits from the Virginia and West Virginia bodies responsible for enforcing the federal Clean Water Act. The SWCB board in Virginia has the authority to deny permits to MVP, which both New York and Connecticut have successfully done [125, 129].[4] However, Virginia's Democratic governors (Terry McAuliffe and Ralph Northam) and staff of the DEQ and other agencies have repeatedly denied that they have this authority.

It is important to note that no local governments can deny permits to MVP. This is far different than the case for renewable energy projects such as wind and solar.

7.2.3 Supporters and Opponents

Opposition and support for MVP do not fall into neat "red" and "blue" categories. Opponents are far more numerous than supporters within 50 miles of the pipeline route and include:

- Landowners and their neighbors
- Republican-dominated county boards of supervisors

---

[3] The AT is a unit of the National Park System, formally designated the Appalachian National Scenic Trail. For simplicity, it will be referred to here as the AT.

[4] Documents originally retrieved online from New York State are no longer in same locations. State's press release: https://www.dec.ny.gov/press/105941.html. State's full letter to applicant: https://www.dec.ny.gov/docs/administration_pdf/constitutionwc42016.pdf. The Constitution Pipeline This project remains embroiled in procedural permit dispute involving the FERC, the DEC and the federal Second District Court of Appeals [Prohaska 136, August 12; 143].

- Gubernatorial primary candidates from both parties
- Members of local, regional, and national conservation, preservation and outdoor groups
- Lawyers from environmental and other organizations
- Activists with experience fighting other pipelines
- Downstream communities impacted by MVP
- Progressive Virginia Democrats
- On procedural issues, both Democratic and Republican members of the U.S. Congress

Opposition is based more on geography than politics. Private property advocates are appalled by the use of federal eminent domain for a project with no local benefit; environmentalists oppose expansion of fossil fuel infrastructure and negative environmental impacts. Landowners and many others, regardless of political affiliation, simply dislike the behavior of MVP and the visible negative impacts to beautiful countryside.

Supporters of the project generally have direct economic ties to it, receive funding from its owners or believe it would benefit the regional or national economy. These include:

- The FERC, which behaves as if MVP is a client rather than an applicant
- Two Democratic governors in Virginia
- The Roanoke Regional Chamber of Commerce
- The Trump White House
- Members of the House and Senate from both parties who receive funding from oil and gas interests and/or support fracking, the use of natural gas in the U.S., and the export of LNG as a strategic counter to Russian influence in Europe and Asia

It is worth noting that many well-informed people within federal, state, and local agencies have worked very hard to do their jobs, although they were often ignored or bullied into submission by key decision makers. Eventually their efforts were vital in overturning deficient federal permits.

7.3 Pipeline Chronology

MVP expected to have their project in service by the end of 2018, probably because they expected little opposition on the physically challenging but sparsely populated route they selected. Instead they have encountered numerous delays due to court challenges and regulatory actions. Meanwhile, the project's parent company, EQT, has seen its stock price plummet and has largely divested its interest, turning the project over to EQM and ETRN, two financially shaky spinoffs of EQT. Both customer demand and LNG prices have dropped, especially in Asia, where at least one MVP owner has a contract to sell LNG. In 2014, Asian prices for LNG were more than six times higher than they were in early 2020 [39].

7.3.1 2014–2015

MVP and the FERC came to town in late 2014, with public briefings and open houses along the route and private briefings with groups such as the Roanoke Regional Chamber of Commerce (EIS pre-filing review process) [46]. MVP published consulting studies on the pipeline's projected economic benefits in December 2014 (FTI Consulting) [45], while the FERC assigned the MVP Docket PF15-3 in their eLibrary[5] and took public comments regarding MVP at six public hearings in early 2015.

Public opposition emerged quickly, with large groups speaking at public hearings, negative resolutions from conservative county boards of supervisors, and unflattering coverage by local media. A loose and diverse coalition of almost two dozen organizations quickly formed and worked in collaboration with opponents of the Atlantic Coast Pipeline.

Hustling to overcome claims that the project provided no local benefits, MVP gained an important supporter in May 2015 after RGC Resources, parent company of local utility Roanoke Gas, became both an owner and a customer. The Roanoke Regional Chamber of Commerce then endorsed the project, assuring the public that MVP would act "in cooperation with property owners, and with the utmost safety and with respect for the environment and our region's beauty" [145].

7.3.2 2016–2017

Although MVP received all of its permits by the end of 2017, it was already running a year behind its ambitious early schedule. These were exhausting years for pipeline opponents, as we had the facts on our side but faced entrenched opposition from powerful national leaders in both major parties.

The deeply deficient DEIS was prematurely released by the FERC [47]. It should have been a complete document for public comment. Instead, the FERC's incoherent review process allowed MVP to add tens of thousands of unindexed pages at any time after publication of the DEIS. The FS, BLM, and other state and federal agencies became cooperating agencies and used the FERC's EIS instead of writing their own. They probably rued this decision, since many of these permits would be withdrawn by the U.S. Fourth Circuit Court of Appeals in 2018 and 2019 due to violations of their own regulations and standards.

---

[5] See https://www.ferc.gov/docs-filing/elibrary.asp for details on how to use this library.

The FERC issued the equally deficient Final Environmental Impact Statement (FEIS) in June 2017, aided by aggressive support from the new Trump Administration, which took an active role in forcing cooperating agencies to accept lower environmental standards than those they originally proposed.

FERC's Commissioners then issued a Certificate of Public Convenience and Necessity on a 2-1 vote in October 2017. Democratic commissioners had always supported pipelines in the past, but opponents of MVP and ACP had convinced Cheryl LaFleur, the only remaining Democratic Commissioner, that companies need to make a stronger case for public need than signing up their own subsidiaries as "customers" [120]. Asked by opponents to provide a rehearing on their permits, the FERC issued a "tolling order" that essentially allows the project to be completed before any decision is made about a rehearing, a procedure that is currently being challenged in federal courts [72].

The FS, BLM, and COE issued their final permits by late 2017. For me, two of the most egregious decisions were those made by the FS. They allowed the AT crossing at a very scenic location Fig. 7.2 and simply lowered the Scenic Integrity Standards of the AT in the Jefferson National Forest. The FS also endorsed exceedingly low erosion and sedimentation/water quality standards for MVP, in direct contradiction to their own earlier filings to the FERC.

At the state level, Virginia's departing Democratic governor, Terry McAuliffe made sure that project would have a free hand. The DEQ negotiated very lax environmental controls along the route and almost nonexistent enforcement of erosion and sedimentation requirements. The City of Roanoke formally objected to likely increases in stormwater costs from MVP sedimentation [1] and opponents mounted a lively campaign, but a divided State Water Control Board (SWCB) issued a permit in December 2017.

Fig. 7.2 Peters Mountain Wilderness from Appalachian Trail, near proposed MVP crossing

7.3.3 2018–2019

2018 started as a very depressing year, with the FERC giving MVP permission to begin cutting trees and a federal court in Roanoke ruling that MVP could begin construction on the property of over 300 unwilling landowners before even agreeing on a price for the easement. During the first half of the year, MVP defeated landowners in one court case after another and proposed a Southgate extension of the pipeline into North Carolina. The dramatic June 7, 2018 explosion of the new, "best in class" Leach Xpress Pipeline in West Virginia confirmed fears that existing "best practices" are inadequate on steep, landslide-prone slopes [98, 99].[6]

Young tree sitters near the AT gave us some hope in the bleak winter months, especially when a local West Virginia judge ruled that they were defenders of the environment. We were also cheered by the national and international coverage of Red and Minor Terry in April 2018.

But the best news came in the summer, when the U.S. Fourth Circuit Court of Appeals pulled MVP's permits from the FS, BLM, and COE. The same court took similar actions against ACP, even ruling that natural gas pipelines cannot cross the Appalachian Trail in national forests without an act of Congress. Opponents began signing emails, "May the Fourth be with you."

By year's end, MVP's estimated costs escalated to $4.6 billion, compared to an initial estimate of $3.0–3.4 billion [100, 104, 108]. EQT, the Pittsburgh fracking company that started the project, was forced to spin off the pipeline part of their company as separately traded EQM, which was seen at the time as financially more stable than the fracking company. In fact, both companies and a second spinoff, ETRN, are now on a downward spiral that has not ended.

MVP continued limited construction in 2019, especially on compressor stations and on the sections in West Virginia and in the flatter Eastern part of Virginia, but negative news continued to mount. MVP is under criminal investigation for possible violations of the Clean Water Act and other federal laws by the U.S. attorney's office in Roanoke and the EPA. A grand jury has been convened [58, 62]. State agencies levied fines of $2.1 million (Virginia) and $266,000 (West Virginia) for hundreds of erosion and sedimentation violations. The company ended the year saying that the earliest completion date would be late 2020, with total costs of $5.0–5.5 billion.

7.3.4 2020

In May 2020, construction on MVP remained stalled, with erosion and sedimentation control the only permitted work in most places. MVP claims the project is 90% complete, but one section in Virginia is less than 20% finished, and another is less

---

[6]Investigators concluded that the Leach Xpress explosion was due to a landslide. It seems likely that both the FERC and the developer knew that this was an entirely unsafe location for a pipeline.

than 50% complete. The company, now largely owned by spinoff ETRN, still says it will be done by the end of 2020.

The first step would be recovery of permits from the FS, BLM, COE and a new Biological Opinion from F&W. The company also awaits the imminent U.S. Supreme Court ruling on whether a permit to cross the AT on federal land requires an act of Congress.[7]

Recovery of MVP's COE permit for crossing wetlands and waterways was complicated by a May 11, 2020 ruling by a federal district court in Montana regarding the Keystone XL pipeline. The judge ruled that COE permits for the Keystone XL Pipeline violate provisions of the Endangered Species Act and specifically stated that the ruling applied to other interstate oil and natural gas pipelines under construction. Height Capital Markets, an investment banking firm, now predicts that MVP is not likely to be completed until the second quarter of 2021 [69].

The impact of the COVID-19 pandemic on MVP's plans is uncertain. Their own FEIS plans suggest that at least 1,200 out of state workers would descend on rural western Virginia for hasty completion of the project. Opponents are already pointing out the dangers of this approach [25].

7.4 Opposition Principles in Action: 2014–2020

Below, I outline my personal list of sound principles for pipeline campaigns, borrowing from and building on civil rights and environmental campaigns of the past. It is important to understand that this is long, hard work requiring patience and endurance. The FERC process is so outrageous that it is easy to get lost in anger and frustration. A few guiding principles can at least channel energy in constructive directions.

7.4.1 Show Up (to Public Meetings and Comment Periods)

We participated fully in several extremely complex regulatory processes, including MVP public briefings and open houses (2014), FERC Scoping sessions (2015), and "public comment" meetings (2016) as well as numerous DEQ and State Water Control Board (SWCB) hearings and meetings (2016–present). Hundreds of opponents often attend such meetings.

We also made thousands of detailed filings to the FERC eLibrary and to other federal agencies, state regulators, and local governments. Between January and November 2015, we wrote most of the 3,800 filings to the FERC eLibrary. We did the same with the Virginia Department of Environmental Quality (DEQ) and West Virginia Department of Environmental Protection (DEP).

---

[7] *United States Forest Service v. Cowpasture River Preservation Association.*

When local landowners or tree sitters ended up in court, we attended the trials. Many of us donated to help pay for the legal fees.

If our pro-pipeline governor comes to town, there will be protesters with signs to greet him. Our volunteers have gone to Pittsburgh, Richmond and Washington, DC to make our views known.

7.4.2 Create a Strong Core and Continually Find New Allies

It is vital to understand that the core of pipeline opposition at the grass roots level cuts entirely across the usual political boundaries. Those with a libertarian streak dislike intrusive federal agencies and the use of eminent domain to seize private property as much as environmentalists despise fossil fuel companies and destruction of ecosystems. Ideological purity is dysfunctional in pipeline fights.

When MVP made their first public presentation to the Roanoke County Board of Supervisors (BOS) in October 2014, I was part of the large crowd in attendance and left the room with a man who had been my opponent over a proposed wind farm in his community several years earlier. He said, "It looks like we're on the same side on this one." We immediately made plans to meet with the chairman of the BOS, an accountant who owns an ice cream shop in my city and is now a Republican representative in our state legislature. We met with him about a week after the MVP presentation, and he immediately told us not to worry because the BOS was going to formally oppose MVP, which they did almost immediately. Similarly, conservative Republican county boards of supervisors in Giles, Craig, and Montgomery counties all voted to oppose MVP by the end of 2015.

Local landowners and environmentalists formed the core of early opposition, and we have continually expanded our geographic and ideological reach. Today, POWHR (powhr.org) is the central anti-MVP organization. They founded Mountain Valley Watch, the citizen group whose trained volunteers are watchdogs on the MVP route, and they play a central role in coordinated filings, expert reports, and legal actions. Below are some examples of how the core expands.

With help from a Sierra Club volunteer who was already working on the Atlantic Coast Pipeline (ACP), opponents representing about two dozen organizations met at Roanoke College in November 2014 and formed a loose coalition. We spread our messages through personal meetings and through websites, email, Facebook groups, and other digital media, assuring that citizens could make public comments, write letters to the editor, and appear at public events. We met several times with a similar group fighting the Atlantic Coast Pipeline (ACP).

Opponents often take reporters, public officials, and others to locations along the MVP route, making it easy to reach remote locations. These visits produce widespread media coverage. Landowners brought in Jane Kleeb, a Nebraskan dubbed the "Keystone Killer" by *Rolling Stone* magazine, to share advice about grassroots activism and plant ceremonial "Seeds of Resistance" on the MVP route [14].

The Grandin Theater Foundation in Roanoke tapped into landowner rights and the improper use of eminent domain by showing the movie, *Little Pink House*, which had been lauded by conservative columnist George Will. The panel that discussed it afterward included a representative of conservative Americans for Prosperity as well as MVP opponents [147].

Like more moderate incumbents, Tea Party Republicans in Craig County with a strong libertarian streak opposed the project in 2016, winning two open supervisory seats on the MVP route. Both candidates posted only two campaign signs in their front yards: "Trump/Pence" and "No MVP." Environmentalists need to remember that rural libertarians are an extremely important part of the effort to defeat projects likes MVP.

The Appalachian Trail Conservancy (ATC) and the Roanoke Appalachian Trail Club (RATC) both worked with the NPS to file strong critical comments to the FERC. The ATC identified MVP as the greatest pipeline threat to the entire trail, advancing these concerns through a Congressional Appalachian Trail Caucus [115]. I was deeply involved in this part of the work, and our studies and simulations on the project's visual impacts to the Trail received widespread attention.

A retired physician and I also worked with Physicians for Social Responsibility to present a forum on energy issues at the Virginia Tech Carilion School of Medicine in Roanoke, highlighting serious potential public health problems of fracking and pipelines. Others have also warned of the public health dangers of pipeline coating left outdoors, exposed to UV rays two years or more [9, 122].

7.4.3 Learn the Facts and Tell the Truth

The truth has been our friend. We conducted research, identified qualified experts, and produced credible reports on the likely impacts of the pipeline as dangerous, unnecessary, and damaging to water, forests, fish, and wildlife. This was often in sharp contrast to much of the work done by MVP. Their inattention to detail was a major reason for project delays. We made thousands of filings to the federal and state agencies and wrote numerous blogs and op eds.

College faculty and retirees were central to this effort. Here are some examples:

- MVP overviews. We held two forums on pipelines at local colleges, Western Virginia Community College in 2015 and Roanoke College in 2016. Both were free, were introduced by senior college officials and included experts on both sides of the issue [26, 27].
- Project need and economics. In addition to publicizing a U.S. Department of Energy study showing no need for MVP or ACP in Virginia or the Carolinas [142], we funded:

- A rebuttal to MVP's case for economic benefits [133].[8]
- A study showing no need for the project [148].
- A study showing how pipeline developers play the FERC's permit system by signing up their own units as "customers," putting communities, ratepayers, landowners, and investors at risk [119].
- Studies estimating that MVP would cost eight counties between $14.5 and $15.3 billion in damages to property values, local tax revenues, tourism, water quality, and other ecosystem services [13, 134].

- Geologic hazards. Over half a dozen reports demonstrated that the MVP route is full of geologic hazards that should be "no-build zones" for pipelines (Shingles). The best-known was written by a retired geology professor and showed that the MVP route risked pipeline explosions due to the simultaneous presence of steep slopes, landslide-prone soils, and unstable karst full of underground streams and caverns. These risks are all magnified by the active Giles County Seismic Zone (GCLC), scene of one of the largest earthquakes in Virginia history [15, 117].
  Another expert report showed that pipeline failures in karst valleys have a minimum safe evacuation distance of 7,544 feet on each side—a swath almost 3 miles wide [138].
- Erosion and sedimentation. The most important study was funded by MVP [42, 43], responding to a FS requirement to assess the project's likely impacts on downstream water quality. The report showed that uncontained MVP erosion would travel far downstream but claimed the effects would be temporary and that the developer could contain 85% of excess sediment. The FERC originally allowed MVP to keep this study secret, but strong protests from experts such as a fisheries science professor at Virginia Tech, resulted in public access to the report in July 2016. FS staff produced a scathing public response to MVP, stating the pipeline would create excess sediment throughout the project's life and could not possibly contain more than 60% of sediment. They chided MVP for using an EPA study conducted in Florida rather than on steep, landslide-prone slopes. These dueling estimates of sediment control would become central in the 2018 decision by the U.S. Fourth Circuit Court to withdraw the FS permit for MVP. The City of Roanoke would use the same study in 2017 to argue that MVP would add $37 million to stormwater cleanup costs in the city every year. In addition, multiple reports by a professional hydrogeologist showed that MVP would have multiple, cumulative, long-lasting negative impacts to all downstream waterways and biological communities, and one of her studies was included in Roanoke County's comments to the FERC [41, 40].
- Visual impacts. MVP and Virginia's governors insisted that the project would have no visual impact because it would be buried underground. I worked with a

---

[8]Phillips had long experience with natural gas pipeline studies. He produced similar studies for Delaware River Keepers and for the opponents of the Atlantic Coast Pipeline, among others. Cardno, the consulting firm hired by FERC to work on MVP, had tried to purchase Key-Log Economics with the proviso that the company would discontinue calculation of indirect costs of pipelines such as water pollution and lost tourism. Private discussion with Phillips.

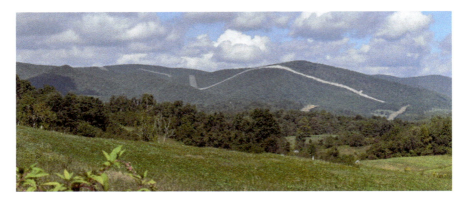

Fig. 7.3 Simulation of MVP impact in Giles County, VA Hill Studio for Roanoke Valley Cool Cities Coalition

graduate of Radford University's geography program to develop maps showing the project's likely impacts and with a landscape architecture firm to develop visual simulations of project impacts Fig. 7.3. These became poster children for MVP's real impact [29, 115 Appalachian Trail Conservancy].

Our work was strongly reflected in DEIS comments by local governments, federal agencies and even a few state agencies. *The Roanoke Times* characterized official responses to the DEIS by saying, "The horse isn't dead but it's surely taking a beating" [6, 7]. The most frequently voiced concerns were the need for a revised or supplemental DEIS with more complete and accurate information about project necessity, alternative routes, economic impacts, impacts on the JNF and the AT, geologic hazards, erosion, and sedimentation control and impacts on plant and animal life, air and water. We successfully made our positions known and credible. The FERC should never have allowed this DEIS to stand.

Even after MVP received its permits, we continued detailed scrutiny and analysis of the project. For example, we showed that pipeline coatings might threaten public health, that MVP's stormwater studies were faulty and that the FERC project manager ignored project requirements to speed up construction just as federal courts were hearing cases about MVP [34, 65, 122, 132].

7.4.4 Talk to Elected Officials and Agency Staff

We met privately with elected officials and staff at all levels of government. This section details examples of our impact.

7.4.4.1 Local Officials

After MVP made their original presentations in 2014, we met privately with county supervisors, who did not need much convincing to vote against the project. By early December 2014, three Republican-dominated county boards (Montgomery, Roanoke, Giles) had voted to oppose MPV. When the route was later changed to include Craig County, their board quickly opposed it as well. Additional meetings resulted in relocation of the MVP route away from a middle-class community and a compressor station out of a county whose citizens opposed it [18].

Two of us worked with Roanoke's city staff and elected officials, and they used MVP's own study [42, 43] to show that the project would add over $37 million annually to stormwater cleanup costs for the City of Roanoke [41]. City staff later helped us prepare documentation for a federal court case that led to the loss of MVP's permit with the U.S. Fish and Wildlife Service in 2018.

7.4.4.2 State Officials

Thanks to strongly pro-pipeline Democratic governors and the strength of utility companies in the state legislature, we have had limited success at the state level. Our own representatives have introduced bills to reduce the power of pipeline companies and strengthen regulation, and they continue to do so. Staff and elected officials from Roanoke met with state officials about erosion issues in the Roanoke River, and the state used MVP mitigation funds to install two new U.S. Geological Survey turbidity gauges upstream from Roanoke.

With our help, Sen. John Edwards (D-Roanoke) personally wrote to DEQ's director in July 2017. He detailed MVP's threats to water quality, enumerated the state's authority and responsibility under the federal Clean Water Act, and repeated constituent concerns that "DEQ is speeding through its review of this important project with inadequate review and public input" [44].

Del. Sam Rasoul (D-Roanoke) used our briefings to focus on MVP's public health and drinking water impacts in a widely covered press conference on the banks of the Roanoke River. Speakers included a local microbrewery owner who expressed concerns about MVP sediment's impact on local beer. Rasoul publicized a map we created showing that MVP would cross waterways over 100 times upstream from the Roanoke metropolitan area and demanded that DEQ examine the full impacts [8, 130, 146]. Rasoul also wrote to state officials asking for more complete studies [137].

When Ralph Northam campaigned for governor in 2017, he privately told us he would hold MVP to higher water quality standards, but he quickly abandoned this position after taking office.

We also worked with the Sierra Club to intervene in a rate case hearing by the State Corporation Commission (SCC) when Roanoke Gas, an MVP customer, increased rates on all customers in January 2019, in part to pay for connections to MVP. The Sierra Club gained legal standing because I swore I am both a Sierra Club member and a Roanoke Gas customer. The involvement of a Sierra Club legal representative prompted the SCC to hire an independent consultant, who contested Roanoke Gas claims of the need for MVP gas. The SCC ultimately denied over half of the rate increase request, including recovery of expenses for connections to MVP.

7.4.4.3 Federal Officials

It has been hard to challenge the FERC and the oil and gas industry, but elected officials of both parties have helped us with procedural issues.

U.S. Sen. Tim Kaine held listening sessions with opponents, and we followed up extensively with his staff. The ATC formed an AT Caucus that includes members from both parties, and they encouraged further study of the need for two pipelines in Virginia. Dissenting FERC commissioners reflected these views during their split votes to approve MVP and rehear the approval.

I worked very closely with NPS and ATC staff on responses to the DEIS and helped arrange a live meeting among MVP, ATC, RATC, NPS, the FS and other federal agencies at Roanoke College in June 2017 to discuss MVP and the Appalachian Trail. ATC staff insisted that MVP would have major visual impacts, and mitigation for these impacts is still under discussion [21]. I addressed the six geologic High Hazard areas identified by the FS in less than four miles of the MVP route in JNF, especially the two areas on each side of the proposed Trail crossing. I asked the MVP construction supervisor to cite one example of a 42-inch natural gas pipeline constructed successfully in an environment that combines steep slopes, landslide-prone soils, karst and an active earthquake zone. His answer was: "FLORIDA" [31].

7.4.5 Find Creative Ways to Protest

We have used creative, fact-based demonstrations throughout the MVP fight, resulting in outstanding media coverage.

In 2014, when MVP held an open house meeting in the Roanoke Valley, there was no opportunity for public comment. We surprised them with a competing event in a room next door, beginning 30 minutes before their event. We had a detailed handout and representatives from over a dozen different organizations, each speaking for only three minutes. Our event garnered far more airtime and print space than MVP's open house [11, 121].

We repeated the tactic in November 2016, when the FERC held a "public" hearing that required private meetings with FERC staff. In addition to our alternative press conference in a nearby auditorium, we posted videos on a new Facebook page [48] showing the private talks between citizens and FERC staff, which Roanoke College students helped produce. Virginia Tech students added lively chants in the waiting room. Our open public hearing of about 200 people made the front page of *The Roanoke Times* [4, 28].

Beginning in 2016, the Sierra Club organized annual fall protests called Hands Across the Appalachian Trail against both MVP and ACP near the pipeline routes. They featured speakers, music, and exhortations to protect the Trail and the communities around it. Students from Roanoke College and Virginia Tech participated [23].

Tree sitting, living on a platform high above the ground on the MVP route, generated enormous media coverage. Tree sitters ranged in age from their 20s to their 70s and included more women than men. Social media posts from tree sitters and their allies included detailed critiques of pipelines and fracking. So far, there have been about a dozen different tree sitters in at least six different locations.

Someone has been sitting on a tree on the MVP route every day since February 26, 2018. The first MVP tree sitters appeared right after the FERC approved tree cutting. They positioned themselves in or near JNF, just downhill from MVP's proposed crossing of the AT. There were soon two more sets of young trees sitters in the JNF. They received minor jail sentences and stalled construction for 3 months while court cases found their way to the U.S. Fourth Circuit [80, 86, 89, 78–93, 101, 60].

The JNF tree sitters inspired a mother and daughter near Roanoke, who climbed trees on their own property in April 2018. Red and Minor Terry thought MVP contractors were violating their federal permits by cutting trees outside an agreed-upon deadline when they showed up with chain saws on the Terry property. Many supporters camped nearby or visited the very accessible site, and their story was a sensation in state and national media, with coverage from *The Washington Post*, *Rolling Stone*, *The Guardian* and many other outlets [81, 128, 135, 139]. Governor Northam ridiculed them.

A young woman from Roanoke County occupied yet another tree on private property near Elliston on September 5, 2018. This Yellow Finch tree sit has been active for over 500 days and has included a 69-year-old grandfather who said he has children and grandchildren to think about [38, 87, 107, 64].

Additional protestors, including a Virginia Tech professor, also began chaining themselves to MVP equipment to stall the work, and over 50 people were arrested for direct action by 2020 [67].

7.4.6 Inform, Involve, and Become the Media

Ongoing positive relationships with the media resulted in hundreds of newspapers and television stories, some reaching far beyond our region, with coverage in *The Washington Post*, *New York Times*, *Los Angeles Times*, *Rolling Stone*, *Guardian* and *Independent*. The pipeline route's obvious destruction, fierce opposition by local residents and successful federal court cases drew the greatest attention.

The best stories can be described in a sound bite, lend themselves to photos and video, and are in locations that the media can easily reach. Landowners near the remote MVP crossing of the Appalachian Trail got reporters to the story using four-wheelers.

The Roanoke Times is clearly the newspaper of record for MVP, posting hundreds of articles since 2014. Reporter Duncan Adams received an award from partners of the AT, and Laurence Hammack was named Reporter of the Year in Virginia [32, 141]. The local CBS affiliate (WDBJ), long-time local ratings leader, often covered the MVP and the NBC (WSLS) and Fox (WFXR) also provided frequent coverage. The leading conservative talk radio station (WFIR) featured interviews with pipeline opponents.

Our location west of the Blue Ridge mountains makes it hard to get media coverage in highly populated eastern markets like Richmond and Washington, DC, but the emergence of new media outlets has enabled us to create news as well as getting others to report it. A typical sequence would be:

1. Opponents report a story on a Facebook page such as Appalachians Against Pipelines.
2. Others quickly share the report on other Facebook pages and social media sites.
3. If the story is meaty enough, someone writes a news article for *Blue Virginia*, an online weekly that focuses on progressive issues, or *The Virginia Mercury*, an independent, nonprofit online newspaper founded by experienced journalists.
4. It is then picked up by *The Roanoke Times*, *The Richmond Times-Dispatch*, or even *The Washington Post* or *Associated Press News*.
5. If the story lends itself to broadcast media (easy to drive to, sound bite and image-ready), it may expand further.

Here are a few specific examples of how we became the media or got them involved in the story.

We drew excellent press coverage with well-organized alternative press conferences at MVP and FERC events in the Roanoke Valley.

We became the media when Marino Colmano launched Lucid Media's 28-video pipeline documentary series and full-length movie titled, *Pipeline Fighters*. The project offers professionally produced short videos with on-site interviews and powerful images of the people, the countryside, and the threats posed by MVP. The feature film received several national and international honors [37].

The ATC created an entire web page devoted to "Mountain Valley Pipeline: a bad energy project approved by a bad energy policy." It featured outstanding short videos and links to media coverage of pipelines for several years (Appalachian Trail Conservancy) [20].

Anti-pipeline signs along a state highway in tiny Newport, Virginia attracted the attention of *The Los Angeles Times*, telling drivers when they were entering and leaving the blast zone and high impact zone of the MVP [53].

An ATC staffer and I broke the story that the FS illegally allowed motorized vehicles on the AT for 19 days, which is strictly prohibited. Armed security from MVP and the FS established a 24 × 7 camp near tree sitters and drove to it instead of walking. We shared still photos with *The Roanoke Times*, which immediately reported, "ATV traffic on the Appalachian Trail is the latest Mountain Valley Pipeline controversy." Embarrassed, the FS apologized a day later [33, 84, 85].

We illustrated the weakness of MVP's plans in May, when the project covered a county highway in up to a foot of mud, resulting in a lot of media coverage and a brief construction suspension [90]. We wrote op eds and blogs about the project's poor design, lack of need and negative impacts [24, 35, 52, 124].

7.4.7 Use the Legal System

Lawsuits focused on two major areas: private property rights and environmental impacts. Landowners face violation of their constitutional rights through MVP trespassing and use of federal eminent domain lawsuits, and pipeline construction poses serious threats to water quality, views from the AT and the survival of endangered species.

7.4.7.1 Private Property Rights

Landowners used local, state, and federal courts to contest MVP's right to enter, survey, and seize the use of their property. They found little support since current laws overwhelmingly endorse the rights of the oil and gas industry rather than those of private citizens.

Shortly after the FERC commissioners issued a Certificate of Public Convenience and Necessity in October 2017, MVP sued over 300 Virginia landowners in Roanoke's federal district court, asking for "immediate access and entry prior to the determination of just compensation" [10]. Known as "quick take," this practice is allowed if the company posts a sufficient bond in court. The developer may then chop down the trees, blast, clear the land and construct the pipeline before agreeing on a price. Opponents packed the courtroom of Judge Elizabeth Dillon when she heard the case on January 11 and 12, 2018. The MVP project manager swore that they needed to start cutting trees by February 1 so that they would be done by March 31, when they would stop for protection of endangered bats and other species. This would prove to be a false statement, as they never stopped cutting trees. On March 2, after MVP produced an acceptable bond, Judge Dillon allowed MVP's quick take, a decision that was upheld on appeal in 2019 [74, 79].

Everyone, including Dillon, believed that MVP would stop cutting trees on April 1. They did not, and a Bent Mountain mother and daughter, Red and Minor Terry, ascended platforms to protect the trees on their property. Minor Terry reports watching MVP biologists RUN through the Terry property, purportedly looking for endangered species, so that the company could continue cutting down trees after the March 31 deadline. The FERC allowed them to do this[9] [72, 74, 75, 77, 79, 61].

MVP asked Judge Dillon to find the Terrys in contempt. When the Terrys' lawyer noted that MVP had claimed they needed immediate access so that they could finish cutting trees by March 31, MVP's lawyers denied they had ever made such a broad statement. Although Dillon ruled in MVP's favor, her decision included a footnote saying she could understand why the Terrys believed that MVP was violating a federal ruling by cutting trees in April. Neither Red nor Minor were fined [87, 88].

Dillon would show little sympathy for MVP when new tree sitters appeared on private property near Roanoke in September 2018. MVP asked her for an injunction, but she directed them to state or local courts [64].

The Terrys did score a victory in a local court. Roanoke County had joined MVP security in placing a 24 × 7 cordon around the Terrys. After they came down, Roanoke County filed criminal charges against them for trespassing (on their own property!), obstruction of justice and interfering with the property rights MVP. The county's lawyer said he wanted to punish Red and Terry by sending them to jail and fining them almost $100,000. The Terry attorney asked whether it was normal for the county to have 24 × 7 armed guards for a misdemeanor charge. He also cited the footnote in Judge Dillon's ruling that said she could understand why the Terrys thought MVP was violating a ruling to stop cutting trees after March 31. Roanoke County General District Court Judge Scott Geddes threw the case out, noting that a criminal charge required a showing that the Terrys intended to break the law, which Geddes seriously doubted [Hammack 70, April 11, April 13, 82, 83, 110, 111].

Overall, though, courts provided almost no relief for landowners. After Virginia landowners lost a 2016 constitutional challenge to a Virginia statute that allows gas pipeline surveyors to enter land without permission, MVP aggressively sent large teams of surveyors onto private lands, and local authorities refused to intervene [16, 17, 3].[10] When the FERC's Commissioners voted 2-1 to authorize MVP in October 2017, landowners appealed to both the U.S. Fourth and Fifth circuits with identical results. The courts told them to go back to the FERC and ask for a rehearing. FERC commissioners voted down the rehearing request on a rare 3-2 split vote, with dissenters questioning the public need for MVP, especially because all the "customers" are part of the companies that own the project [73, 76, 78, 95, 59, 61]. In January 2020, landowners launched a new constitutional challenge to the Natural Gas Act of 1938 in the U.S. Fifth Circuit, but the court once again allowed the FERC to go unchecked [51, 57, 69].

---

[9]Comments by Minor Terry at public showing of "The Little Pink House" at the Grandin Theater in Roanoke. We were both on a post-film panel.

[10]The relevant statute in Virginia, 56-49.01, does not require the companies to demonstrate public use [12].

7.4.7.2 Environmental Impacts: "May the Fourth Be with You" and the Lorax

We won an early victory in 2018 when a local judge refused to remove the first set of tree sitters near the AT, saying there was no urgent need for natural gas and that tree sitters "generally represent the interest of the public and the environment" [116].

Meanwhile lawyers from the Sierra Club, Appalachian Mountain Advocates and other groups brought successful suits before the U.S. Fourth Circuit Court of Appeals in Richmond. The availability of legal resources from these environmental groups was invaluable. In June 2018, the court stayed MVP's authority to construct stream crossings in West Virginia under the Clean Water Act, and a successful petition from the Sierra Club and three other conservation groups extended the stay to include crossings in Virginia. In November, the same court completely withdrew MVP's authorization under Section 404 of the Clean Water Act, and the company still lacks these permits from the COE [96, 97, 106, 109, 54]. In late July, the court struck down permits from the FS and the BLM that allowed MVP to cross the JNF and the AT [102].

The same Fourth Circuit ruled in December that the FS does not have the authority to allow the Atlantic Coast Pipeline (ACP) to cross the AT. This case is referred to as the "Lorax" case because Judge Stephanie Thacker quoted from Dr. Seuss, saying, "We trust the United States Forest Service to 'speak for the trees, for the trees have no tongues.'" The appeal on this complex case was heard by the U.S. Supreme Court in February 2020. If it stands, no natural gas pipelines will be allowed to cross the Trail on federal property without an act of Congress, and MVP would probably need to change their route. A ruling is expected in June 2020, with most observers predicting that the court will overturn the Fourth Circuit and allow the Forest Service to issue permits for AT crossings [22, 144].

The FERC's goal seems to be allowing continuation of construction regardless of how many federal permits MVP lost in court. For example, in May the FERC began issuing variances to get around the pipeline's loss of authorizations from the COE. The FERC project manager essentially allowed the entire project to become a variance, rejecting advice from his own experts [34].

More trouble for MVP came after rainfall in July 2019 resulted in major releases of mud into the Roanoke River. In August, the Sierra Club filed suit against the U.S. Fish & Wildlife Service (F&W) in the Fourth Circuit, showing that MVP was violating the Endangered Species Act (ESA) by failing to protect at least four species of fish and bats. Many of us provided the documentation, and I am a declarant in the case. MVP and the FERC tried to stall court action by appearing to stop work. After complaints about FERC's refusal to stop work from several county governments and 20 state representatives and senators, the Fourth Circuit finally stepped in, staying the F&W permit. MVP cannot legally continue work without it, and the project remains officially idle except for erosion and sedimentation control in May 2020 [66, 114]. Earlier in the project, FERC had repeatedly allowed MVP to continue construction and pipe installation in the name of "erosion control," and pipeline opponents remain ever-vigilant.

We had poor results with the Commonwealth of Virginia on environmental issues. The Fourth Circuit rejected our appeal of the Virginia's SWCB's water permit for MVP. Unlike the FS, the DEQ and the SWCB never disputed the flimsy erosion and sedimentation control plans offered by MVP. This made it difficult for courts, which they are loath to say they are hydrology experts [103]. After citizens of Mountain Valley Watch (MVW) documented hundreds of MVP erosion and sedimentation violations that were ignored by the DEQ, Attorney General Mark Herring used the MVW data to sue MVP in December 2018. He signed a Consent Order with MVP in October 2019, assessing a relatively small fine and allowing MVP to police itself by hiring someone to monitor the route [112, 113].

7.4.8 Be Watchdogs on the Pipeline Route

Local residents were always the best sources of information about MVP's behavior. Their efforts resulted in media coverage, fines, stop work orders and a lawsuit from the Commonwealth of Virginia. After the project received its permits in 2017, POWHR formed Mountain Valley Watch (MVW), a largely volunteer effort by residents, landowners, and concerned citizens; engineering and scientific experts; nonprofit organizations and environmental groups, and Virginia Tech students. Trained MVW observers perform drone, fixed wing and on-the-ground monitoring, with the results documented, entered into a GIS data base and reported to the DEQ, SWCB and others [94, 127].

Dozens of citizens received training, and any time it rained, they grabbed their cameras and went to the pipeline route. They conducted two Vioblitzes in 2019, documenting violations along the entire MVP route in both Virginia and West Virginia during a single weekend.

When state and federal regulators failed to stop the continually failing project, MVW submitted a 34-page report for the August 2018 SWCB meeting. It was full of photos and detailed reports [105]. The SWCB surprised the DEQ, the Attorney General, and the Governor by voting to reconsider the MVP permit in December 2018, but they were told this was legally risky and took no further action. The state government appears more worried about a lawsuit from MVP than about water quality and property rights in Virginia [55, 56, 63].

7.4.9 Stay the Course

The opposition has not gone away. Instead, deep friendships have been formed while our predictions of pipeline impacts have sadly been confirmed on the ground. On signs and in correspondence opponents say "WANGA," meaning "We Are Not Going Away."

Shortly after the 2016 Presidential election, opponents of both MVP and ACP held a summit at a Natural Bridge hotel sponsored by two coalitions, POWHR and the Allegheny-Blue Ridge Alliance. About 180 people attended. Joe Lovett, executive director of Appalachian Mountain Advocates, summed up the situation when he said, "Our job just got much harder." He also accurately noted that the inaccuracies of MVP's DEIS could enhance efforts to halt the project through litigation. Nebraskan Jane Kleeb, a leader in the fight against the Keystone XL pipeline, urged participants to continue building a diverse coalition in defense of private property rights and clean water. Chase Iron Eyes, a tribal leader from the Standing Rock Sioux Reservation in North Dakota, elicited a standing ovation. "Stay strong," he said. "That's the only chance we have" [5].

In 2019, the POWHR Coalition and Mountain Valley Watch were honored with the Appalachian Studies Association's Helen M. Lewis Community Service Award and e-Appalachia Award, respectively. Our work continues on all fronts. The POWHR Coalition Newsletter in January 2020 summarized events of 2019 and listed new community programming in 2020, including a local speaker series on climate change, sustainability, and clean energy; individual and community-level options for solar and an Environment at Risk Art Show in Richmond.

7.5 Conclusions

When MVP came to the Roanoke Valley, they assured everyone that there was great need for their product. The opposite has proven true. As *The New York Times* recently reported, "A glut of cheap natural gas is wreaking havoc on the energy industry, and companies are shutting down drilling rigs, filing for bankruptcy protection and slashing the value of shale fields they had acquired in recent years" [118]. Energy giant, Chevron, is writing down $10 billion to $11 billion in assets, mostly shale gas holdings in Appalachia and a planned liquefied natural gas export facility in Canada [118].

MVP has not resumed construction at this writing, with many legal and financial challenges unresolved. The company hopes to regain the federal permits that it has lost from the FS, F&W, BLM, and COE in 2020. It is unclear exactly how these processes will play out, and they could cause additional significant delays and potential additional lawsuits.

EQT, the Pittsburgh fracking company that started MVP, is selling its interest in the project to help reduce its massive debts and rescue its bonds from a junk rating [36]. Its spinoff successors, EQM and ETRN, have also seen low stock prices and face an international glut of LNG, plummeting LNG prices in Asia and many investor questions about MVP.

Two court cases loom large. The U.S. Supreme Court will soon decide whether the FS has the authority to issue permits for natural gas pipeline crossings of the AT. If the Fourth Circuit is upheld, MVP will probably have to change the crossing of the AT in the JNF. In the second case, a federal District judge in Montana held in

May 2020 that the COE must adhere to provisions of the ESA in granting permits for Keystone XL and other new pipeline projects to cross waterways and wetlands. This decision could add a year of more of delay and additional costs for MVP [68].

Projects like MVP will threaten all Americans until the FERC approval process is radically reformed and the Natural Gas Act of 1938 is replaced or reinterpreted. Nor will real change occur until politicians from both parties reject natural gas as a "bridge" fuel to renewable energy. It is time to move on.

References

1. Adams, D (2017, September 5) City officials discuss pipeline's impact on Roanoke River. The Roanoke Times. Retrieved from https://www.roanoke.com/
2. Adams D (2014, October 14) Mountain Valley Pipeline plan raises ire in Roanoke County. The Roanoke Times. Retrieved from https://www.roanoke.com/
3. Adams D (2016, October 18) Bent Mountain property owners seek injunction to bar surveying. The Roanoke Times. Retrieved from https://www.roanoke.com/
4. Adams D (2016, November 3) Pipeline watchdogs slam analysis of proposed project's environmental impacts. The Roanoke Times. Retrieved from https://www.roanoke.com/
5. Adams D (2016, November 12) Pipeline opponents' summit emphasizes resolve and unity in wake of Trump's election. The Roanoke Times. Retrieved from https://www.roanoke.com/
6. Adams D (2016, December 20) Regional governments bash pipeline impact statement. The Roanoke Times. Retrieved from https://www.roanoke.com/
7. Adams D (2016, December 21) EPA finds fault with environmental review of Mountain Valley Pipeline. The Roanoke Times. Retrieved from https://www.roanoke.com/
8. Adams D (2017, July 18) DEQ agrees to add informal meetings on Mountain Valley Pipeline. The Roanoke Times. Retrieved from https://www.roanoke.com/
9. Adams D (2017, October 27) Forum highlights perils of fossil fuels, pipelines and feeling disempowered. The Roanoke Times. Retrieved from https://www.roanoke.com/
10. Adams D (2017, October 27) Mountain Valley sues landowners to gain pipeline easements and access through eminent domain. The Roanoke Times. Retrieved from https://www.roanoke.com/
11. Adams D (2014, December 18) Salem pipeline open house an example of conviction, unity of opponents. The Roanoke Times. Retrieved from https://www.roanoke.com/
12. Adams D (2015, June 25) Pipeline turnabout: gas could be sent to India. The Roanoke Times. Retrieved from https://www.roanoke.com/
13. Adams D (2016, May 18) Study backed by Mountain Valley Pipeline opponents suggests negative economic impacts for region. The Roanoke Times. Retrieved from https://www.roanoke.com/
14. Adams D (2016, June 7) 'Seeds of Resistance' tour aims to repel pipeline with planting of sacred corn. The Roanoke Times. Retrieved from https://www.roanoke.com/
15. Adams D (2016, July 7) Mountainous karst landscape should be a 'no build' zone for pipeline, geologist says. The Roanoke Times. Retrieved from https://www.roanoke.com/
16. Adams D (2016, October 12) Pipeline surveying conflict heats up on Bent Mountain. The Roanoke Times. Retrieved from https://www.roanoke.com/
17. Adams D (2016, October 17) Roanoke Gas proposes tap in Montgomery County for Mountain Valley Pipeline. The Roanoke Times. Retrieved from https://www.roanoke.com/
18. Adams D (2016, October 17) Roanoke County police will not defy state law allowing pipeline surveying without permission. The Roanoke Times. Retrieved from https://www.roanoke.com/

19. Alinsky S (1971) Rules for radicals: a pragmatic primer for realistic radicals. Vintage Books, New York
20. Appalachian Trail Conservancy (ATC) Mountain Valley Pipeline: a prime example of bad energy development. Retrieved from https://www.appalachiantrail.org/home/conservation/advocacy/mountain-valley-pipeline
21. Appalachian Trail Conservancy (ATC) (2017, February 22) Response to MVP visual impact study. FERC e-Library, Docket CP16-10, Accession No. 20170222-5062
22. Atlantic Coast Pipeline LLC v. Cowpasture River Preservation Association. Retrieved from https://www.scotusblog.com/case-files/cases/united-states-forest-service-v-cowpasture-river-preservation-association/.
23. Blue Virginia (2019, September 29) Hundreds celebrate fourth annual "Hands Across the Appalachian Trail." Blue Virginia. Retrieved from https://bluevirginia.us/2019/09/hundreds-celebrate-fourth-annual-hands-across-the-appalachian-trail
24. Bondurant R, Leech I (2019, November 21) The 'public need' argument for the MVP grows weaker. The Roanoke Times. Retrieved from https://www.roanoke.com/opinion/commentary/bondurant-and-leech-the-public-need-argument-for-the-mvp/article_c139b078-7918-5d5c-ae2b-b405b89ae5c5.html
25. Bowers K, Christopulos D, Smusz T (2020, April 1) COVID-19 and an army of pipeline workers don't mix. Retrieved from The Roanoke Times at https://www.roanoke.com/opinion/commentary/bowers-christopulos-and-smusz-covid-19-and-an-army-of-pipeline-workers-dont-mix/article_aff886d5-4f60-5da2-ae45-b3df75e2b7be.html
26. Christopulos D (2015, March 3) Final agenda for March 9 Forum on Natural Gas Pipelines—registration is free! Retrieved from https://rvccc.org/2015/03/03/final-agenda-for-march-9-forum-on-natural-gas-pipelines-registration-is-free/
27. Christopulos D (2016, October 16) Presentations from October 12, 2016 Forum on Natural Gas Pipelines. Retrieved from https://rvccc.org/2016/10/16/presentations-from-october-12-2016-forum-on-natural-gas-pipelines/
28. Christopulos D (2016, November 26) Troubling use of police power at November 3 FERC meeting in Roanoke. Retrieved from https://rvccc.org/2016/11/26/troubling-use-of-police-power-at-november-3-ferc-meeting-in-roanoke/
29. Christopulos D (2017, January 1) Virginia's Blue Ridge or Virginia's pipeline alley? The Roanoke Times. Retrieved from: https://www.roanoke.com/
30. Christopulos D (2017, June 15) Notes from meeting on MVP. Contemporaneous personal notes
31. Christopulos D (2017, September 18) Earthquakes and pipelines: recipe for disaster. Retrieved from https://www.ratc.org/earthquakes-and-pipelines-recipe-for-disaster/
32. Christopulos D (2017, November 15) Graceful farewell from Duncan Adams, Roanoke Times reporter. Retrieved from https://www.ratc.org/graceful-farewell-from-duncan-adams-roanoke-times-reporter/
33. Christopulos D (2018, May 4) Mountain Valley Pipeline drove ATVs on the Appalachian Trail for 19 days. Retrieved from https://www.ratc.org/mountain-valley-pipeline-drove-atvs-on-the-appalachian-trail-for-19-days/
34. Christopulos D (2019, January 22) New proof: entire Mountain Valley Pipeline project based on known falsehoods. Retrieved from https://www.ratc.org/new-proof-entire-mountain-valley-pipeline-project-based-on-known-falsehoods/
35. Christopulos D (2019, June 19) Good thing Martin Luther King didn't rely on The Roanoke Times for strategy. Retrieved from https://www.roanoke.com/opinion/commentary/christopulos-good-thing-martin-luther-king-didn-t-rely-on/article_dd678cb3-cd3d-566a-8c81-c1ba2d0762a7.html
36. Cocklin J (2018, November 13) Retrieved from https://www.naturalgasintel.com/articles/116456-eqt-completes-midstream-spinoff-equitrans-trading-on-nyse
37. Colmano M (2016) Retrieved from https://pipelinedocumentary.com/
38. Dashiell J (2020, January 17) Tree-sitters mark another milestone in Montgomery County. Retrieved from https://www.wdbj7.com/content/news/Tree-sitters-mark-another-milestone-in-pipeline-protest-567094901.html

39. Dezember R (2020, February 26) Shale gas swamps Asia, pushing LNG prices to record lows. Wall Street Journal
40. Dodds P (2016, December 22) Hydrological assessment of watershed impacts caused by constructing MVP through Roanoke County, Virginia. Contained in FERC Accession No. 20161222-5459
41. D'Ardennes D, Weitzenfeldt L (2017, September 25) Mountain Valley Pipeline: risks for the city of Roanoke
42. ESI (2016, June 7) Hydrologic analysis of sedimentation. Prepared for: U.S. Department of Agriculture, Forest Service, Jefferson National Forest, Eastern Divide Ranger District. Prepared on behalf of Mountain Valley Pipeline. Retrieved from FERC e-Library, Docket No. CP16-10
43. ESI (2017, March 3) Revised hydrologic analysis of sedimentation. Retrieved from FERC e-Library, Docket No. CP16-10. Accession No. 20170303-5014. Prepared for: U.S. Department of Agriculture, Forest Service, Jefferson National Forest, Eastern Divide Ranger District. Prepared on behalf of Mountain Valley Pipeline
44. Edwards J to Paylor D (2017, July 27) DEQ and MVP. Personal copy of letter
45. FTI Consulting (2014, December 10) Economic benefits of Mountain Valley Pipeline Project in Virginia. For Mountain Valley Pipeline
46. Federal Energy Regulatory Commission (FERC) EIS pre-filing review process. Retrieved from https://www.ferc.gov/resources/processes/flow/process-eis.asp
47. Federal Energy Regulatory Commission (FERC) (2016, September) Draft Environmental Impact Statement (DEIS). Mountain Valley Project and Equitrans Expansion Project. FERC Docket Nos.: CP16-10-000 and CP16-13-000
48. Federal Energy Regulatory Commission (FERC) (2016, November 3) Transcript. Scoping [sic] meeting in Roanoke, Va. Regarding Mountain Valley Pipeline LLC. CP16-10
49. Federal Energy Regulatory Commission (FERC) (2017, June) Final Environmental Impact Statement (FEIS). Mountain Valley Project and Equitrans Expansion Project. FERC Docket Nos.: CP16-10-000 and CP16-13-000
50. Federal Energy Regulatory Commission (FERC) (2017, October) Order Issuing Certificates and Granting Abandonment Authority re Mountain Valley Pipeline. FERC Docket Nos.: CP16-10-000 and CP16-13-000. Accession No. 20171013-4002
51. Gentry-Locke Attorneys (2020, January 3) Starting off the New Year with a 'Bang!': Virginia Landowners quietly file constitutional case against FERC
52. Hadwin T (2019, September 23) Roanoke Gas customers will pay for MVP. The Roanoke Times. Retrieved from https://www.roanoke.com/opinion/commentary/hadwin-roanoke-gas-customers-will-pay-for-mvp/article_61ca26e6-9cab-54c6-9b56-7218bb52e70c.html
53. Halper E (2017, July 14) A pipeline cutting through the iconic Appalachian Trail sparks a fight over natural gas expansion. Los Angeles Times. Retrieved from https://www.latimes.com/
54. Hammack L (2018, November 27) Mountain Valley Pipeline loses another water-crossing permit. The Roanoke Times. Retrieved from https://www.roanoke.com/
55. Hammack L (2018, December 7) Virginia files lawsuit against Mountain Valley Pipeline. The Roanoke Times. Retrieved from https://www.roanoke.com/
56. Hammack L (2018, December 13) State board to reconsider key permit for Mountain Valley Pipeline. The Roanoke Times. Retrieved from https://www.roanoke.com/
57. Hammack L (2019, January 4) Latest lawsuit against Mountain Valley Pipeline contests the taking of private land. The Roanoke Times. Retrieved from https://www.roanoke.com/
58. Hammack L (2019, January 16) Mountain Valley Pipeline starts the new year with new complications. The Roanoke Times. Retrieved from https://www.roanoke.com/
59. Hammack L (2019, January 22) Roanoke attorneys seek criminal investigation of Mountain Valley Pipeline. The Roanoke Times. Retrieved from https://www.roanoke.com/
60. Hammack L (2019, January 23) Pipeline protester sentenced to 14 days in jail. The Roanoke Times. Retrieved from https://www.roanoke.com/

61. Hammack L (2019, February 5) Appeals court allows quick-take of land for Mountain Valley Pipeline. The Roanoke Times. Retrieved from https://www.roanoke.com/
62. Hammack L (2019, February 15) Criminal investigation of Mountain Valley Pipeline underway, document shows. The Roanoke Times. Retrieved from https://www.roanoke.com/
63. Hammack L (2019, February 28) MVP asks state board to discontinue process aimed at stopping pipeline construction. The Roanoke Times. Retrieved from https://www.roanoke.com/
64. Hammack L (2019, April 4) After 212 days, tree-sitters are still standing against the Mountain Valley Pipeline. The Roanoke Times. Retrieved from https://www.roanoke.com/
65. Hammack L (2019, July 31) Pipe coating is safe, Mountain Valley tells regulators. The Roanoke Times. Retrieved from https://www.roanoke.com/
66. Hammack L (2019, September 19) Plans for pipeline work area atop Poor Mountain draw objections from Roanoke County. The Roanoke Times. Retrieved from https://www.roanoke.com/
67. Hammack L (2020, January 30) Charge dismissed against Mountain Valley Pipeline opponent. The Roanoke Times. Retrieved from https://www.roanoke.com/
68. Hammack L (2020, May 12) Federal judge upholds ban on process for permitting pipelines, including Mountain Valley Pipeline. The Roanoke Times. Retrieved from https://www.roanoke.com/
69. Hammack L (2020, May 14) Judge dismisses lawsuit that contested Mountain Valley's power of eminent domain. Te Roanoke Timesh. Retrieved from https://www.roanoke.com/
70. Hammack L (2018, April 11) Police make arrests as protests of the Mountain Valley Pipeline intensify
71. Hammack L (2018, April 13) Roanoke County cuts off supplies to tree-sitters blocking pipeline crews
72. Hammack L (2017, December 12) Judge narrows lawsuit over efforts to take private land for Mountain Valley Pipeline. The Roanoke Times. Retrieved from https://www.roanoke.com/
73. Hammack L (2018, January 9) Environmental groups seek stay to halt construction of Mountain Valley Pipeline. The Roanoke Times. Retrieved from https://www.roanoke.com/
74. Hammack L (2018, January 12) Mountain Valley begins legal efforts to take private land for its pipeline. The Roanoke Times. Retrieved from https://www.roanoke.com/
75. Hammack L (2018, January 13) Efforts to take land for the Mountain Valley Pipeline challenged by property owners. The Roanoke Times. Retrieved from https://www.roanoke.com/
76. Hammack L (2018, January 28) Environmental groups attack federal approval of Mountain Valley Pipeline. The Roanoke Times. Retrieved from https://www.roanoke.com/
77. Hammack L (2018, January 31) Federal judge puts a pause on Mountain Valley Pipeline construction plans. The Roanoke Times. Retrieved from https://www.roanoke.com/
78. Hammack L (2018, February 5) Federal appeals court declines to block construction of Mountain Valley Pipeline. The Roanoke Times. Retrieved from https://www.roanoke.com/
79. Hammack L (2018, March 5) Judge allows Mountain Valley Pipeline work to proceed on private property. The Roanoke Times. Retrieved from https://www.roanoke.com/
80. Hammack L (2018, March 28) Tree-sit protest of Mountain Valley Pipeline escalates, drawing police response. The Roanoke Times. Retrieved from https://www.roanoke.com/
81. Hammack L (2018, April 2) A tree-sit protest of the Mountain Valley Pipeline has spread to Roanoke County. The Roanoke Times. Retrieved from https://www.roanoke.com/
82. Hammack L (2018, April 19) Roanoke County police charge 2 women in trees blocking the Mountain Valley Pipeline. The Roanoke Times. Retrieved from https://www.roanoke.com/
83. Hammack L (2018, April 23) Roanoke County cuts off supplies to tree-sitters blocking pipeline crews. The Roanoke Times. Retrieved from https://www.roanoke.com/
84. Hammack L (2018, April 30) ATV traffic on the Appalachian Trail is the latest Mountain Valley Pipeline controversy. The Roanoke Times. Retrieved from https://www.roanoke.com/
85. Hammack L (2018, May 1) Forest Service apologizes for damage to Appalachian Trail during patrols of pipeline protests. The Roanoke Times. Retrieved from https://www.roanoke.com/

86. Hammack L (2018, May 2) Road closure violates the free-speech rights of pipeline protesters, lawsuit claims. The Roanoke Times. Retrieved from https://www.roanoke.com/
87. Hammack L (2018, May 4) Judge finds 'Red' Terry and her daughter in contempt for tree-sit protests of pipeline
88. Hammack L (2018, May 5) Protesters leaving tree-stands on Bent Mountain after being found in contempt of court. The Roanoke Times. Retrieved from https://www.roanoke.com/
89. Hammack L (2018, May 16) Lawsuit questions treatment of Mountain Valley Pipeline protester. The Roanoke Times. Retrieved from https://www.roanoke.com/
90. Hammack L (2018, May 20) Construction halted at Mountain Valley Pipeline work site following severe erosion in Franklin County. The Roanoke Times. Retrieved from https://www.roanoke.com/
91. Hammack L (2018, May 21) Another Mountain Valley Pipeline protest is off the ground in Giles County. The Roanoke Times. Retrieved from https://www.roanoke.com/
92. Hammack L (2018, May 23) Pipeline protester known as 'Nutty' has come down from her pole in Giles County. The Roanoke Times. Retrieved from https://www.roanoke.com/
93. Hammack L (2018, June 1) Final 2 pipeline protesters are gone from aerial perches. The Roanoke Times. Retrieved from https://www.roanoke.com/
94. Hammack L (2018, June 2) Environmental watchdogs: a citizens' group monitors the Mountain Valley Pipeline. The Roanoke Times. Retrieved from https://www.roanoke.com/
95. Hammack L (2018, June 15) FERC upholds approval of Mountain Valley Pipeline project. The Roanoke Times. Retrieved from https://www.roanoke.com/
96. Hammack L (2018, June 21) Appeals court issues stay of permit for Mountain Valley Pipeline in W.Va. The Roanoke Times. Retrieved from https://www.roanoke.com/
97. Hammack L (2018, June 26) Mountain Valley Pipeline foes file new legal challenge following last week's win. The Roanoke Times. Retrieved from https://www.roanoke.com/
98. Hammack L (2018, July 18) Pipeline explosion in W.Va. cited by opponents of Mountain Valley Pipeline. The Roanoke Times. Retrieved from https://www.roanoke.com/
99. Hammack L (2018, July 23) Environmental regulators cite Mountain Valley Pipeline again. The Roanoke Times. Retrieved from https://www.roanoke.com/
100. Hammack L (2018, July 25) Completion of Mountain Valley Pipeline delayed to early 2019, even with long work days. The Roanoke Times. Retrieved from https://www.roanoke.com/
101. Hammack L (2018, July 26) Protester gets 2 days in jail for blocking construction of the Mountain Valley Pipeline. The Roanoke Times. Retrieved from https://www.roanoke.com/
102. Hammack L (2018, July 27) Federal appeals court delivers blow to Mountain Valley Pipeline. The Roanoke Times. Retrieved from https://www.roanoke.com/
103. Hammack L (2018, August 1) Appeals court upholds Virginia's review of water quality impact of Mountain Valley Pipeline. The Roanoke Times. Retrieved from https://www.roanoke.com/
104. Hammack L (2018, August 17) Mountain Valley Pipeline cuts workforce, delays project completion to late 2019. The Roanoke Times. Retrieved from https://www.roanoke.com/
105. Hammack L (2018, August 18) As state board takes up pipeline permits, thousands of comments await. The Roanoke Times. Retrieved from https://www.roanoke.com/
106. Hammack L (2018, August 30) Federal court allows stream crossing work for Mountain Valley Pipeline. The Roanoke Times. Retrieved from https://www.roanoke.com/
107. Hammack L (2018, September 5) Pipeline protesters take to the trees near Elliston. The Roanoke Times. Retrieved from https://www.roanoke.com/
108. Hammack L (2018, September 25) Estimated cost of Mountain Valley Pipeline increased to $4.6 billion. The Roanoke Times. Retrieved from https://www.roanoke.com/
109. Hammack L (2018, October 5) Stream-crossing permit suspended for Mountain Valley Pipeline in Virginia. The Roanoke Times. Retrieved from https://www.roanoke.com/
110. Hammack L (2018, November 8) Charges heard against Roanoke County mother and daughter who sat in trees to block a pipeline. The Roanoke Times. Retrieved from https://www.roanoke.com/
111. Hammack L (2018, November 15) Judge dismisses charges against Roanoke County women who sat in trees to block pipeline. The Roanoke Times. Retrieved from https://www.roanoke.com/

112. Herring M (2018, December 7) Attorney General Herring and DEQ file lawsuit over repeated environmental violations during construction of Mountain Valley Pipeline. Retrieved from https://www.oag.state.va.us/media-center/news-releases/1341-december-7-2018-herring-and-deq-file-suit-over-environmental-violations-during-construction-of-mountain-valley-pipeline?highlight=WyJtb3VudGFpbiIsInZhbGxleSIsInBpcGVsaW5lIiwicGlwZWxpbmUncyIsIm1vdW50YWluIHZhbGxleSIsIm1vdW50YWluIHZhbGxleSBwaXBlbGluZSIsInZhbGxleSBwaXBlbGluZSJd
113. Herring M (2019, October 11) MVP, LLC to pay more than $2 million, submit to court-ordered compliance and enhanced, independent, third-party environmental monitoring. Retrieved from https://www.oag.state.va.us/media-center/news-releases/1548-october-11-2019-mvp-llc-to-pay-more-than-2-million-submit-to-court-ordered-compliance-and-enhanced-independent-third-party-environmental-monitoring?highlight=WyJtb3VudGFpbiIsInZhbGxleSIsInBpcGVsaW5lIiwicGlwZWxpbmUncyIsIm1vdW50YWluIHZhbGxleSIsIm1vdW50YWluIHZhbGxleSBwaXBlbGluZSIsInZhbGxleSBwaXBlbGluZSJd
114. Hurst C to Bose K (2019, October 1) Re: Docket No. CP16-10-000 (Mountain Valley Pipeline)
115. Igelman J (2017) Cutting to the core. AT Journeys. Retrieved from https://www.ratc.org/
116. Irons R (2018, March 27) Mountain Valley Pipeline, LLC vs Appalachians Against Pipelines, et al. Circuit Court of Monroe County, West Virginia. Action No. CC-32-2018-C-2
117. Kastning E (2016, July 3) An expert report on geologic hazards in karst regions of Virginia and West Virginia: investigations and analysis concerning the proposed Mountain Valley Gas pipeline. Retrieved from https://wp.vasierraclub.org/KastningReport.pdf
118. Krauss C (2019, December 11) Natural gas boom fizzles as a U.S. glut sinks profits. New York Times. Retrieved from https://www.nytimes.com/2019/12/11/business/energy-environment/natural-gas-shale-chevron.html
119. Kunkel C, Sanzillo T (2016, April) Risks associated with natural gas pipeline expansion in Appalachia. Institute for Energy Economics and Financial Analysis
120. LaFleur C (2017, October 13) Dissent on order issuing certificates and granting abandonment authority to Mountain Valley and Atlantic Coast pipelines. Retrieved from https://www.ferc.gov/media/statements-speeches/lafleur/2017/10-13-17-lafleur.asp#.XtAKSMZ7mL8
121. Lambertsen K (2014, December 18) Pipeline opponents question report: are there any benefits? WSLS 10 News. Retrieved from https://www.wsls.com/story/27656616/pipeline-opponents-question-report-are-there-any-benefits
122. Limpert W (2019, June 30) Pipeline coating is dangerous. Retrieved from https://www.roanoke.com/opinion/commentary/limpert-pipeline-coating-is-dangerous/article_abc3409c-291a-5ab3-ab5a-ab2214e82061.html
123. Main I (2016, July 10) Southeastern electric utilities find their way to higher profits through gas pipelines and captive consumers. Power for the People. Retrieved on May 15, 2020 https://powerforthepeopleva.com/2016/07/10/southeastern-electric-utilities-find-their-way-to-higher-profits-through-gas-pipelines-and-captive-consumers/
124. Majors L, Chisholm R, Bondurant R (2018, March 14) Majors, Chisholm and Bondurant: a dispatch from the path of the Mountain Valley Pipeline. Retrieved from https://www.roanoke.com/opinion/commentary/majors-chisholm-and-bondurant-a-dispatch-from-the-path-of/article_1b9203e5-f26f-5c1f-9948-87a9169a796a.html
125. McCarthy G to Muhlherr G (2006, December 19) Water Quality Certification Application #200300937-SJ. Retrieved from https://www.ct.gov/deep/lib/deep/declaratory_rulings_other_decisions/islandereastdecision.pdf.
126. Moody's Investor Service (2020, April 2) Moody's downgrades EQT to Ba3; outlook remains negative. Retrieved from https://www.moodys.com/research/Moodys-downgrades-EQT-to-Ba3-outlook-remains-negative--PR_422015
127. Mountain Valley Watch. Retrieved from https://powhr.org/mvwatch/
128. Nobel J (2018, May 15) Pipeline protesters take to the trees. Rolling Stone. Retrieved from https://www.rollingstone.com/politics/politics-news/pipeline-protesters-take-to-the-trees-628691/

129. Ong K (2018, May 1) Supreme Court lets NY denial of Constitution Pipeline stand. Retrieved from https://www.nrdc.org/experts/kimberly-ong/supreme-court-lets-ny-denial-constitution-pipeline-stand
130. Ostarly C (2017, July 18) Local brewery concerned about Mountain Valley Pipeline. Retrieved from https://www.wfxrtv.com/news/local-news/local-brewery-concerned-about-mountain-valley-pipeline/
131. POWHR (Protect Our Water, Heritage, Rights). Retrieved from https://powhr.org/
132. POWHR (2020, January) Recap of major events in 2019. POWHR Coalition Newsletter
133. Phillips S (2015, October 6) Reason for caution: Mountain Valley Pipeline economic studies overestimate benefits, downplay costs. Key-Log Economics
134. Phillips S, Wang S, Bottorff C (2016, May) Economic costs of the Mountain Valley Pipeline. Key-Log Economics
135. Philp G (2018, May 26) America's tree sitters risk lives on the front line. Retrieved from https://www.theguardian.com/environment/2018/may/26/tree-sitters-appalachian-oil-pipeline-virginia-west
136. Prohaska T (2019, August 12) DEC rejects National Fuel's Northern Access Pipeline – again. The Buffalo News. Retrieved from https://buffalonews.com/2019/08/12/dec-rejects-national-fuels-northern-access-pipeline-again/
137. Rasoul S to Havens M, Paylor D, Dunn R (2017, July 18) Comments on Mountain Valley Pipeline Additional 401 Water Quality Conditions. Personal copy
138. Rubin P (2016, December 2) Expert Report of Paul A. Rubin on behalf of Giles and Roanoke Counties, Virginia. FERC Accession Nos. 20161222-5458 and 20161222-5459
139. Schneider G (2018, April 1) Perched on a platform high in a tree, a 61-year-old woman fights a gas pipeline. The Washington Post. Retrieved from https://www.washingtonpost.com/local/virginia-politics/perched-on-a-platform-high-in-a-tree-a-61-year-old-woman-fights-a-gas-pipeline/2018/04/21/3b8284b4-435e-11e8-bba2-0976a82b05a2_story.html
140. Seidel D (2019, October 11) Virginia, MVP announce agreement for stricter monitoring. Radio IQ/WVTF. Retrieved from Mountain Valley Pipeline, LLC vs.Appalachians Against Pipelines, et al. Retrieved from https://www.wvtf.org/post/virginia-mvp-announce-agreement-stricter-monitoring#stream/0
141. The Roanoke Times (2019, April 6) Yancey, Hammack lead Roanoke Times state press winners. Retrieved from https://www.roanoke.com/news/local/yancey-hammack-lead-roanoke-times-state-press-winners/article_7771a18d-f6e6-517e-a1c2-9bfaf8510edc.html
142. US Department of Energy (DOE) (2015, February) Natural gas infrastructure implications of increased demand from the electric power sector. Retrieved from https://www.energy.gov/sites/prod/files/2015/02/f19/DOE%20Report%20Natural%20Gas%20Infrastructure%20V_02-02.pdf
143. Walton R (2019, September 3) Delayed since 2016, Constitution Pipeline scores win on New York water permit. Retrieved from https://www.utilitydive.com/news/delayed-since-2016-constitution-pipeline-scored-win-on-new-york-water-perm/562110/
144. Wamsley L (2018, December 14) Quoting 'The Lorax,' court pulls permit for pipeline crossing Appalachian Trail. National Public Radio. Retrieved from https://www.npr.org/2018/12/14/676950106/quoting-the-lorax-court-pulls-permit-for-pipeline-crossing-appalachian-trail
145. Waugh J (2015, June 2) The pipeline offers economic opportunities. The Roanoke Times. Retrieved from https://www.roanoke.com/opinion/commentary/waugh-the-pipeline-offers-economic-opportunities/article_c14bbc5c-6bf4-5e71-98e3-a5174ace0b42.html
146. Wickline A (2017, July 18) Roanoke community expresses concern about pipeline's potential impact on drinking water. Retrieved from https://www.wsls.com/news/virginia/roanoke/roanoke-community-expresses-concern-about-pipelines-potential-impact-on-drinking-water.
147. Will G (2018, April 18) Hollywood's newest action star: the Constitution's takings clause. The Washington Post. Retrieved from https://www.washingtonpost.com/opinions/hollywoods-newest-action-star-the-constitutions-taking-clause/2018/04/18/1d7ae45c-4264-11e8-ad8f-27a8c409298b_story.html

148. Wilson R, Fields S, Knight P, McGee E, Ong W (2016, September 12) Are the Atlantic Coast Pipeline and the Mountain Valley Pipeline necessary? Prepared for the Southern Environmental Law Center and Appalachian Mountain Advocates by Synapse Energy Economics, Inc.

Chapter 8
Linking Sovereignty, Local Environments, and Climate Justice Through Pipeline Pedagogy

Theodor Gordon, Corrie Grosse, and Brigid Mark

Abstract Native nations are on the frontlines of resisting pipelines and at the forefront of creating the kinds of relationships required for living well in a just future. Drawing on our experiences working with Native nations in Minnesota as well as teaching, learning, and organizing around North American pipelines, we argue for a pipeline pedagogy that centers sovereignty by supporting Native leaders. Community partnerships, activism, and college courses are three methods for advancing this pipeline pedagogy, which connects environmental and climate justice, social movements, and Native nation sovereignty. We reflect on our experiences in these three domains: developing partnerships between a tribal school and our institutions, teaching and learning in environmental studies courses at a liberal arts college, and organizing with community activists against the Line 3 tar sands pipeline. We offer suggestions for approaches and activities that advance a pedagogy with the power to protect people, prevent pipelines, and promote justice.

Keywords Climate justice · Sovereignty · Indigenous pedagogy · Settler colonialism · Line 3 pipeline · Activism · Tar sands

Native communities have long been on the frontlines of injustice. They faced genocide, forced assimilation through boarding schools, bans on their spirituality, and removal from their homelands. Today, settler colonialism—the practice of one society seeking to move permanently onto a place with which another society is already in deep relation, a place that is intimately their home—continues to reinforce these oppressions [37]. Normalized by a history of treaty violations, North American

T. Gordon · C. Grosse (✉) · B. Mark
College of Saint Benedict and Saint John's University, Collegeville, MN, USA
e-mail: cgrosse001@csbsju.edu

T. Gordon
e-mail: tgordon@csbsju.edu

B. Mark
e-mail: brigid.mark@colorado.edu

pipelines are settler colonialism [37]. Our positionality is as white settlers living in Ojibwe and Dakota territory, teaching and learning at white settler institutions.

A form of settler colonialism, The Dakota Access Pipeline (DAPL), a fracked oil pipeline, threatens water that holds cultural significance to those who inhabit, and whose ancestors inhabited, the land. The DAPL crosses underneath Lake Oahe, the dammed portion of the Missouri River where it meets the Cannonball River, a confluence that once created sacred stones. As LaDonna Brave Bull Allard [7] recounts:

> The stones are not created anymore, ever since the US Army Corps of Engineers dredged the mouth of the Cannonball River and flooded the area in the late 1950s as they finished the Oahe dam. They killed a portion of our sacred river. I was a young girl when the floods came and desecrated our burial sites and Sundance grounds. Our people are in that water. This river holds the story of my entire life.

A spill would contaminate the drinking water of the Standing Rock Sioux, fundamentally violating their sense of identity and connection to the land and waters of their homeland while jeopardizing their survival and well-being. The river also provides drinking water for eighteen million people downstream. Understanding Indigenous peoples' relationship to place is critical for pipeline pedagogy because it identifies the disruption of eco-social relations as violence ([4 pp. 5–6]) that threatens all people and supports resurgence of Indigenous communities [32].

Leanne Betasamosake Simpson, who is Michi Saagiig Nishnaabeg and a member of Alderville First Nation in what is called Ontario, describes the place-based way of knowing and living embodied by Brave Bull Allard's story above as Nishnaabewin, or Nishnaabeg intelligence.[1] According to Simpson [33], Nishnaabeg intelligence signifies:

> the comingling of emotional and intellectual knowledge combined in motion or movement, and the making and remaking of the world in a generative fashion within Indigenous bodies that are engaged in accountable relationships with other beings (p. 21).

Three elements of Nishnaabeg intelligence are central to the pipeline pedagogy we prioritize: (1) engagement with affective and cognitive ways of knowing; (2) responsibility through accountable relationships with people and the more-than-human world, including place; and (3) Indigenous-led creation of a just and healthy world.

Approaching pedagogy from multiple ways of knowing is both consistent with Indigenous forms of pedagogy and central to social justice education. Simpson's "Land as Pedagogy: Nishnaabeg Intelligence and Rebellious Transformation" [32] argues for theory, generated from relationships, movement, and processes, and connection to homeland, as vital for supporting Indigenous leaders. Social justice education's goals are for students to analyze systems of oppression, understand their

---

[1] Glen Coulthard and Simpson [9] call this place-based way of knowing "grounded normativity." They explain that grounded normativity "houses and reproduces the practices and procedures, based on deep reciprocity, that are inherently informed by an intimate relationship to place," and "teaches us how to live our lives in relation to other people and nonhuman life forms in a profoundly nonauthoritarian, nondominating, nonexploitative manner."

own positionality in those systems, and act to change those systems in the interest of social justice ([1], [8]). Learning activities developed by educators in this area engage both emotion and intellect to achieve these goals [1]. Our pipeline pedagogy, therefore, seeks to create experiential, relational, and action-based learning environments.

Responsibility, relationships, and accountability are central to many Indigenous life-ways and essential for good engagements among Native and non-Native individuals, educators, and researchers ([3], [25], [33], [34], [38]). These values are also central to feminist pedagogies and methodologies (e.g. [19]) and to the ways that diverse communities work together to resist fossil fuels ([18, 28]). As non-Native educators, activists, and learners, we seek to affirm and practice these values inside and outside the classroom.

Finally, the pipeline pedagogy we are developing is one that prioritizes Indigenous leadership. As thinkers and doers in many fields committed to social justice stress, the people who are most affected by injustice are best equipped for identifying the roots of injustice and building sustainable and just solutions to systems of oppression (e.g. [12], [28], [36]). The notion of thinking and acting for the good of seven generations ahead, communicated in the original teachings of different Indigenous communities in North America, is one example of a practice that, if followed by all people, would fundamentally change our engagement with each other, the land, and pipelines.

Drawing on our experiences working with Native nations in Minnesota as well as teaching, learning, and organizing around North American pipelines, this chapter presents a pipeline pedagogy that centers sovereignty by supporting Native leaders. We reflect on our experiences in three domains: developing partnerships between a tribal school and our institutions, teaching and learning in environmental studies courses at a liberal arts college, and organizing with community activists against the Line 3 tar sands pipeline. In each setting, we present methods for helping students understand connections between environmental and climate justice, social movements, and Native nation sovereignty; identify their own positionalities in the systems of oppression and privilege that characterize these themes; and build the skills for action to resist oppression, be good allies, and create a world that protects people, prevents pipelines, and promotes justice.

8.1 Native Leaders and Direct Partnership (Theodor Gordon)

Because Native nations are at both the forefront of the negative impacts of pipeline development and the forefront of the activism against their development, strategies for developing effective Native leadership are also strategies for resistance to pipelines. In this section, we propose that pipeline pedagogy intersects with Native pedagogy and we reflect on the steps we are taking to link the two, including our partnership with Nay-Ah-Shing, a school operated by the Mille Lacs Band of Ojibwe.

While education has the potential to develop future leaders in all social arenas (e.g. education, politics, movements), it has also been deployed by the United States as a tool of settler colonialism. Before we could take steps toward partnering with Nay-Ah-Shing, we needed to investigate our (the authors') own institutions' (The College of Saint Benedict and Saint John's University, from here on CSBSJU) role in implementing assimilationist pedagogies against the Ojibwe. First, this section provides a discussion of CSBSJU's past complicity with assimilationist policies. The second part describes our partnership with Nay-Ah-Shing and how it led us to link Native and pipeline pedagogies.

For over a century, the United States pursued a policy of forced assimilation that required the separation of Native children from their families by placing them in boarding schools [40]. These assimilationist policies enlisted both church and military institutions in the settler colonialist goal of removing Native nations from their land. Among the hundreds of Native American Boarding Schools, two were operated on the current campuses of CSBSJU, located in central Minnesota. From 1884 to 1896, St. Benedict's Monastery operated the St. Benedict's Industrial School and St. John's Abbey operated the St. John's Industrial School. At their peak, these schools each enrolled over 100 Ojibwe students, with funding from and a curriculum developed by the federal government.

Today, Native communities across the United States experience significant educational disparities that are directly connected to past and present education policies that marginalize Native youth ([30], [39]). All educational institutions have a responsibility to be inclusive of Native youth in ways that respond to past assimilationist policies. Because of its specific involvement in implementing these policies, we believe that CSBSJU has a particular obligation to serve Native youth today in the terms defined by their own communities. Many universities have been complicit in historical injustices, from slavery and segregation to assimilation. If CSBSJU can successfully develop a partnership that combines Native and pipeline pedagogies in response to our institutional histories, then we believe we may serve as a model for other universities working to redress their own historical injustices. Below, I describe and reflect on our steps to develop this partnership.

While some histories of our campuses include references to the industrial schools, most community members at CSBSJU have been unaware of this past. As a cultural anthropologist and a contingent faculty member, I was interested in researching the schools but colleagues had advised me not to out of concern that unearthing painful parts of our campuses' past could jeopardize my career. Fortunately, in Fall 2017 CSBSJU released a call for proposals for faculty research projects to make our campuses more inclusive. I successfully applied for a grant to investigate the past and present experiences of Native students on our campuses. With institutional support, I began archival research into the industrial schools in Spring 2018.

In Summer 2018, I attended the Native Studies Summer Workshop for Educators, hosted by the Mille Lacs Band. At the workshop, I met with faculty and administrators at Nay-Ah-Shing. I offered to share research results as they develop, initiating the first sharing of materials on CSBSJU's industrial school past with Native communities. Our discussion also included the current needs of the students at Nay-Ah-Shing. In particular, their faculty and administrators noted a need to increase college attendance as well as the need to further develop Nay-Ah-Shing's focus on project-based environmental pedagogy. Nay-Ah-Shing is located a 75-minute drive from CSBSJU. Because CSBSJU offers outdoor educational programming, an Environmental Studies department, and enrolls between 20 and 30 Native students, we saw the potential for CSBSJU to serve these needs.

After our meeting, I followed up with Nay-Ah-Shing to ask for specific guidance about how to address the needs they identified. They were particularly interested in our Outdoor U program, which provides K-12 outdoor educational programming on our campuses, which include over 3,000 acres of forest, prairie, and lakes. Activities are led by student naturalists and supervised by staff. Nay-Ah-Shing specifically requested content that focused on sustainability. I worked with Outdoor U to identify activities focused on invasive species that addressed sustainability and included hands-on activities in our forest. Nay-Ah-Shing also identified the need for their students to experience a college campus and meet students in an informal setting. We recruited CSBSJU students to volunteer to pair up and join the Nay-Ah-Shing students for lunch during their visit. Many of the volunteers were education students, eager to work with youth from diverse backgrounds.

Once we arranged for these activities, Nay-Ah-Shing asked that we prepare their students for their visit by meeting them ahead of time. One week before their visit, an Outdoor U student naturalist, Kateri Heymans, and I traveled to Nay-Ah-Shing for classroom visits to meet the students and share materials to prepare them for their upcoming visit. I quickly discovered that my experience in college classrooms did not translate well to the middle and high school. Fortunately, Kateri is experienced working with K-12 students and quickly got them engaged and excited for their upcoming visit. In October 2018, 32 students and 8 staff from Nay-Ah-Shing traveled to SJU for the activities we planned. Their students and staff expressed gratitude and we agreed to plan future activities.

When we debriefed with them, their faculty expressed interest in connecting with faculty who could teach the science of climate change and the movements to stop it. They identified a specific need to help their students understand the dangers posed by pipelines. Having listened to their requests, we (Corrie and Theodor) paired with Christen Strollo (an Associate Professor of Chemistry) to develop lesson plans to address these needs.

In January 2019, we (Theodor, Corrie, and Christen), visited middle school students at Nay-Ah-Shing during their science class. First, Christen led a climate science activity that got the students up and moving. They played a game she developed where students took on the roles of molecules and light waves interacting in our atmosphere, simulating the warming effects of greenhouse gases. Then, Corrie led a discussion of the different causes and negative impacts of greenhouse gas pollution. This featured her recent visit to the United Nations Climate Summit in Poland and highlighted the work of youth and Native activists at the summit. Corrie tied this to pipeline development, specifically the Line 3 pipeline that Mille Lacs, and other Native nations in Minnesota, are resisting. Our goal was for students to understand the threats posed by climate change and pipeline development and to know that they can become leaders in the movement to stop these threats.

During this visit, some of the students asked, "where is Kateri?"—she was unable to be at our January visit. Hearing them ask for her led me to realize the impact that she and other volunteers had, but also the limitations of one-time connections. In order to establish meaningful mentorships, our students need to have repeated connections with their students. However, this is inherently challenging, given our students' busy schedules. Before we left, we met with faculty from Nay-Ah-Shing to plan a visit during the Spring semester. They shared that their high school students would benefit from mentorship by current Native students at CSBSJU. Their hope was that by connecting with Native college students, their high school students might be more likely to see themselves as having the potential to succeed in college. They wanted their middle school students to learn about student-led initiatives to make CSBSJU more sustainable, in order to inspire their students to create their own initiatives at Mille Lacs.

In April 2019, Nay-Ah-Shing returned for their second visit of the school year. Two of our (CSBSJU) students led their middle school students through activities that introduced them to sustainability initiatives on our campuses. Four of our Native and Indigenous students took their high school students on a campus tour and held an informal discussion of their experiences as college students. Some of our students shared social media contacts with their students, opening up the potential for them to stay in touch after their visit. We are now in conversations with Nay-Ah-Shing to develop methods to assess the outcomes of our partnership and to apply for grants to grow it.

During the 1880s and 1890s, federal policies removed Ojibwe youth from their homes and brought them to CSBSJU and other institutions for assimilation. One hundred twenty years later, Ojibwe youth are returning. This time, it is to serve the needs identified by their community. With our faculty expertise in climate change science and pipeline activism we are developing activities that help their students learn how to respond to environmental threats. Connecting our resources with their needs is just the start. Our partnership is now growing through a reflective and iterative process. Each activity is followed by a debriefing of its successes and the points that need to be strengthened, leading to the plans for the next activity.

In this way, Native pedagogy can intersect with pipeline pedagogy. We found that in the course of developing a university/Native nation partnership, needs for pipeline pedagogy emerged alongside needs for developing future Native leaders. By building partnerships that connect university resources with the needs of Native youth, pipeline pedagogy can be part of a toolkit that proactively trains the next generation of Native leaders. Repairing relations not only consists of building bridges between Native nations and the university, but also transforming education within the university, addressing the long history of omission and misrepresentation of Native issues by centering pipeline pedagogy.

8.2 Classroom Strategies (Corrie Grosse)

As a sociologist teaching environmental studies courses at a liberal arts college, I have the privilege of employing pipeline pedagogy in many of my courses. Here, I describe classroom activities, assignments, and orientations that support the kind of pipeline pedagogy we prioritize: one that emphasizes connections and uplifting diverse voices while helping students to identify their positionality and to creatively envision a just future. My classes tend to be majority white students, many from the middle class. Therefore, instructors with different student demographics should consider how their students will engage with the following activities.

8.2.1 Understanding Connections

My mentor, John Foran, defines sociology as the study of everything [15], saying that the first principle of sociology is that everything is interconnected. He then points out that this is the same core principle of ecology and Buddhism. As an educator trained in this thinking, I resonate with Native theories and explanations which emphasize interconnections, such as Simpson [33], who explains that cultural and political practices are "joined and inseparable" (p. 49) and generated in relationship to place.

In my teaching, research, and activism, I center climate justice because of its emphasis on interconnections. In contrast to the larger climate movement, the climate justice movement engages in activism for radical social change, recognizing that a livable world can only emerge through social justice. Climate justice presents a constant opportunity to focus on what meaningful involvement (a core element of the US Environmental Protection Agency's definition of environmental justice) in environmental decision-making looks like and how centering the most marginalized communities will uplift everyone (a foundational insight of Black feminist theory (e.g. [22]). Climate justice is a valuable way to move students who are alarmed by what climate change will mean for their own lives toward thinking about the root causes of climate change, its uneven impacts, and where radical solutions lie—solutions geared toward system change, solutions long practiced by Native peoples.

On the second day of my Energy and Society upper-division course, I set the tone with an examination of the mobilization at Standing Rock. To understand the connections between systematic injustices made visible by Standing Rock, we learn about the stories of different participants: the youth, the women leaders [2], and those affected by the militarization at the camps [14]. The stories of youth often resonate particularly well with the students; we hear how youth started the mobilization as well as how they used the community they built to support each other during the ongoing trauma of Native youth suicide [13].[2] We then read an excerpt from Pope Francis'

---

[2]After teaching with this text, I was able to ask Jasilyn Charger, the youth leader featured in the article, what she thought of it. She critiqued the author's intrusion into Jasilyn's personal life and generally said that stories don't always get told in her words, as she would like. I have used this

encyclical *On Care for Our Common Home* [20], which fits with our institutions' Benedictine Catholic traditions. Students also learn the latest news in fossil fuel divestment, a systematic change-making strategy in which they can participate.

I find these materials effective for helping students understand that any question of energy, environment, climate, or infrastructure is tied up with much bigger questions about social (in)justice. I emphasize that different people experience and respond to social injustice in different ways, with young people and women often at the forefront of resistance and solutions. Centering social (in)justice from day one is important for encouraging empathy and critical thinking about the economic and political agendas driving energy. It ensures that students are asking feminist questions like *Who benefits? Who loses?* and *Who is most affected?* about realities and proposed solutions to environmental problems.

I also use film to bring pipeline struggles to life and center the voices of Native participants, leaders, and filmmakers (see *Awake: A Dream from Standing Rock* [16], *The Eagle and The Condor: From Standing Rock with Love* [24], *First Daughter and the Black Snake* [31]). Near the end of my course, we focus on the Alberta tar sands and the Keystone XL pipeline. Students read excerpts from *A Line in the Tar Sands* [6], an anthology with contributions from Indigenous and allied activists and academics. I employ this text over others because of the diverse voices it uplifts.

The tar sands are a good focal point because they illuminate the vastness of climate justice implications around pipelines, connecting communities in struggle across large geographical and cultural distance. Tar sands extraction violates the sovereignty of First Nations in what is called Canada while the shipping of tar sands-processing machinery [17] and the pumping of the oil through pipelines violates treaty rights of Native nations in what is called the United States. Tar sands are also the most greenhouse-gas-intensive form of oil on the planet; through climate change, the burning of tar sands violates treaties again, by affecting the plants and animals upon which Native communities depend. In our state of Minnesota, the proposed Line 3 tar sands pipeline, if built, would jeopardize *manoomin* (wild rice), a center point of Ojibwe culture, subsistence, and economics. The study of pipelines is powerful because it unveils interconnections of social injustices and environmental degradation.

8.2.2 Identifying Positionality

Beyond building knowledge, a pipeline pedagogy must help students identify and grapple with their positionality. Upon finishing *A Line in the Tar Sands*, students in my course write an essay in which they consider if and how they would engage in resistance to the Keystone XL pipeline under the conditions that (1) they were invited to do so by a friend, (2) all costs are paid, and (3) they would not miss

---

as a teachable moment to discuss research ethics with students and how the media manipulates information.

any work or school. I ask them to examine what would motivate or deter them, or other people, from participating. Even as a hypothetical situation, this prompt produces thoughtful reflections that help students connect their own lives, privilege, feelings, and knowledge to pipelines. In Spring 2019, this prompt was particularly real. Then President Trump tried to fast-track the Keystone XL pipeline, prompting a few students and I to attend the Promise to Protect Training (https://nokxlpromise.org/) in preparation to resist.

I also facilitate examination of positionality through an environmental privilege worksheet adapted from Gregory Mengel [10], building on McIntosh [27]. The worksheet asks students a series of questions like, "I can enjoy National Parks like Yosemite and Yellowstone, imagining them as intact wildernesses because their establishment did not involve the forcible removal of my ancestors." If students answer "Yes," they add one point. It also includes questions that highlight all aspects of social identities (class, race, disability, sexuality, and gender). Students fill out the sheet individually and debrief in discussion; many (especially rural Minnesotan students who have grown up surrounded by woods and lakes) have never considered environmental dimensions of privilege and how it differs for white, Native, and other peoples of color. The individual nature of the activity and presentation of a handful of optional discussion questions (*How did you feel doing this?*, *What did it feel like to have a low/high score?*, *What items had you never thought about?*, *Would you give some of your privilege to others?*, *What did you learn?*) protects the privacy of students; they can choose to reveal their answers during discussion or not. Through activities like these, students gain awareness of positionality and privilege—an important first step for engaging in solidarity work.

8.2.3 Cultivating Skills for Action

Whenever possible, I encourage students to participate in events and build relationships outside the classroom. I brought students to the Promise to Protect Training mentioned above; the Rise for Climate, Jobs, and Justice Summit (https://riseforclimate.org/); and the Rising Voices conference on connecting Indigenous and western science to address climate change (https://risingvoices.ucar.edu/). I engage student research assistants in solidarity work with Native-led Honor the Earth (http://www.honorearth.org/); in Summer 2019, at the suggestion of Honor the Earth's Executive Director Winona LaDuke, my research assistant created a water protector welcome packet for water protectors arriving in Minnesota to resist the Line 3 pipeline (see https://www.welcomewaterprotectors.com). For final research projects, I require students to research change-making efforts of organizations and to interview their members.

Inside the classroom, I employ a number of methods for building skills necessary for leadership and for being good allies. To build empathy, I use role-plays where students research and present (in character) the views of diverse stakeholders in pipeline and other energy struggles. Students role-playing decision-makers must then come to a conclusion about whether to approve or deny energy infrastructure. I assign students to work in groups where members have different life experiences and strengths. To create efficient and inclusive discussion, I teach students discussion protocols and hand signals used by Occupy Wall Street—something I learned during my own engagement in climate justice activism. Finally, I facilitate discussions around imagination and envisioning what a world of *buen vivir*,[3] or living well, would entail. I task students with envisioning a zero-carbon day in their future and creatively communicating that vision. Students have produced creative writing, short films, visual art, and choreographed dances.

Imagining a bright future and connecting with social movements building solutions are important for empowering students with the hope to move forward. As Rebecca Solnit [35] writes, "Hope locates itself in the premises that we don't know what will happen and that in the spaciousness of uncertainty is room to act [...]" (p. xiv).

A classroom pipeline pedagogy should emphasize interconnections, help students position themselves within systems of power and privilege, and equip students with skills, empathy, imagination, and passion to be effective leaders and allies to Native communities on the frontlines of pipeline resistance. At its best, this classroom pipeline pedagogy inspires students to seek out their own community engagement opportunities where they become activists, support Native leadership, and develop their own pedagogy. Below, Brigid Mark describes her student experience of pipeline pedagogy.

8.3 Activism (Brigid Mark)

As I progressed through environmental studies courses at my college, it became increasingly isolating to learn about the enormity of climate change and social injustices. As Aldo Leopold [26] eloquently articulates "one of the penalties of an ecological education is that one lives alone in a world of wounds." I dutifully became vegetarian, recycled, and took short showers. But these prescribed individual solutions did not alleviate my sense of helplessness.

Then, I enrolled in Energy and Society, a course described in the previous section, with Professor Corrie Grosse. To learn that two issues of great importance to me, environmental degradation and social injustices, are inextricably interconnected, profoundly affected me. One powerful example shared by *A Line in the Tar Sands* [5] is the connection between the sexual violence against Indigenous women perpetrated by men living in "man-camps" brought in for pipeline construction, and the violence against the earth represented by the pipelines themselves. That environmental and social issues could be addressed simultaneously through system change gave me hope. This class compelled me to participate in activism.

---

[3] *Buen vivir* is a concept enshrined in the Ecuadorian and Bolivian constitutions. It is in opposition to the capitalist notion of living better, with success defined by consumption and profit.

Professor Grosse introduced me to Minnesota 350, a climate justice organization, and I joined the pipeline resistance team. With this team, I helped organize the "Block (Line 3) Party," in May of 2017. The event resisted the proposed Line 3 tar sands pipeline which would expedite climate change, poison water and ecosystems, and violate the treaty rights of the Anishinaabe, a group of Indigenous peoples living in what is called Canada and the upper Midwest United States, including northern Minnesota.

Activism deepened my understanding of three concepts introduced in the Energy and Society course which are essential to our pipeline pedagogy: hope, privilege, and centering the voices of those most affected by injustices. One particular moment at the block party paints a picture of what hope through activism looked like for me:

The odd gathering which was the Block (Line 3) Party, with art and music and eating and laughing, seemed quite out of place in downtown Saint Paul, where suits and briefcases walked in strict lines following strict schedules. The proud outline of the teepee juxtaposed the harsh, blocky figure of the Minnesota Public Utilities Commission (PUC) building, the agency tasked with making the final decision on the pipeline. Painted in bold letters on the teepee's canvas was the message "Hey PUC, Deny Line 3." Some had just returned from a march to the capitol building to deliver a petition to the governor. Some snacked on a community meal composed of traditional wild rice and other typical dishes, warm from the kitchens of Indigenous grandmothers. We were listening to a music performance; then it began to rain. Some ran for cover, but the majority stayed out in the rain. And then they began to dance.

There was so much hope and beauty and solidarity in that moment. It was a temporary reclamation of stolen land, a challenge to the world to laugh, and a firm request for systematic change.

In addition to hope, activism gave me a deeper understanding of privilege. Two years earlier, I took a class called the Social Construction of Whiteness. *The History of White People*, [29] shocked me by exposing how my ancestors could not have lived here, on the very land upon which I now stand, without the invention of race in order to justify the domination of others. As I followed Debby Irving's [23] journey through *Waking Up White*, I realized that my family's middle-class status results partially from the GI Bill, which gave my grandfather housing loans and scholarships to pursue an education, privileges denied to so many others. Irving also helped me understand that a phase I went through as a child, dressing up as a Native American and worshipping their "noble" culture, thoroughly obscured the bloody history of genocide against Indigenous peoples [23].

Activism changed my privilege from something I thought about, to something I felt. For example, before, I thought that pipeline resistance meant camping alongside water protectors obstructing the pipeline route. Through activism, I learned it would be more helpful if I provided the funds and resources for someone else with a more personal connection to the land and who otherwise would not be able to camp. Activism continues to facilitate the uncomfortable and necessary process of realizing my own privilege, a vital part of promoting Native sovereignty.

Finally, activism emphasizes the necessity of centering the voices of those most affected by injustices. Through my activism against the Line 3 pipeline, I had a chance to practice putting this idea into action. During the Block (Line 3) Party, Bill Paulson, an Indigenous community leader, spoke about the poverty he sees every day, and the future of pipelines and undrinkable water he hopes will never come to be. I located a microphone and measured the space so his stage would fit. For Native grandmothers cooking a community meal, I located a kitchen and pots and pans. Since I have profited from the same history and institutions which facilitated the dispossession and poverty of Native peoples, it felt right for me to reject the role of a protagonist, taking a supporting, behind-the-scenes role through my activism. I attempted to reverse the process of stealing and silencing by giving and listening. I have much to learn from powerful Native voices.

When I brought my learning from my activism back to campus, sharing with my peers, they too were compelled by the cause, particularly when framing the issue as not only an environmental threat, but also a social injustice. Our student organization, Climate Action Club, initiated a divestment campaign, created events on campus to raise awareness about the Line 3 pipeline, and integrated social justice elements into our meetings and club structure. Following *Organizing Cools the Planet* [28], we implemented a horizontal leadership structure, began meetings with a land acknowledgment, and included preferred gender pronouns in our introductions. In 2020, we changed our name to Climate Justice Club.

Participating in activism deepened my understanding of interconnections, hope, and privilege, while allowing me to build accountable relationships with those most affected by environmental injustices. This is key to supporting sovereignty and promoting an Indigenous-led creation of a just and healthy world. Using my knowledge from the classroom to inform my activism with Native communities and my peers enables me to not only understand the intersectional injustice of pipelines, but to also feel and act on that injustice.

Integrating activism as a life-giving tool is essential to teaching and learning about pipelines. It is not enough to learn about injustices centered around pipelines, we have to offer a path to follow and the hope to continue.

8.4 Conclusion (All)

From Native nations, to the classroom, to the streets, we see generative spaces for engaging pipeline pedagogy. In all of these contexts, we create and welcome multiple ways of learning and knowing, center the goal of building reciprocal and accountable relationships among people and the more-than-human world, and prioritize leadership by folks in our region on the frontlines of pipeline injustice—Native nations.

The goal of our pedagogy, to borrow from Kimberlé Crenshaw [11] "should be to facilitate the inclusion of marginalized groups, for whom it can be said: 'When they enter, we all enter'" (p. 167). We are putting this into practice in multiple ways. By creating opportunities for a tribal school to integrate university expertise in sustainability into their curriculum, while supporting Native middle and high school student connections to Native and non-Native college students, we are working to support future Native leaders. By centering Native voices in the classroom and educating our students on interconnections, positionality, and change-making through the lens of climate justice, we are cultivating leadership in Native students and building non-Native allies. Finally, by practicing activism on and off-campus, recognizing privilege, and building relationships with Native peoples, we are building movements necessary for system change. We agree with Simpson [33], who writes, "we need to join together in a rebellion of love, persistence, commitment, and profound caring and create constellations of co-resistance, working together toward a radical alternative present based on deep reciprocity and the gorgeous generative refusal of colonial recognition" (p. 9). In the spirit of building constellations of co-resistance, we will strive to cultivate connections with the many knowledge-holders resisting pipelines around the world. This work is critical to survival of the people, cultures, and places that make life possible and fulfilling. As the Native-led grassroots organization leading pipeline resistance in our territory puts it:

> Pipelines threaten the culture, way of life, and physical survival of the Ojibwe people. Where there is wild rice, there are Anishinaabeg, and where there are Anishinaabeg, there is wild rice. It is our sacred food. Without it we will die. It's that simple.

– Honor the Earth "Treaty Rights and Oil Pipelines" fact sheet [21]

References

1. Adams M, Bell LA (eds) (2016) Teaching for diversity and social justice, 3rd edn. Routledge, New York and London
2. Arasim E, Orielle Lake O (2016, October 29) 15 Indigenous women on the frontlines of the Dakota Access Pipeline resistance. EcoWatch. Retrieved from https://www.ecowatch.com/indigenous-women-dakota-access-pipeline-2069613663.html
3. Bacon JM (2017) 'A lot of catching up', knowledge gaps and emotions in the development of a tactical collective identity among students participating in solidarity with the Winnemem Wintu. Settler Colonial Studies 7(4):441–455. https://doi.org/10.1080/2201473X.2016.1244030
4. Bacon JM (2018) Settler colonialism as eco-social structure and the production of colonial ecological violence. Environmental Sociology, 1–11. https://doi.org/10.1080/23251042.2018.1474725
5. Black T, D'Arcy S, Weis T, Russell JK (2014) Introduction. In: Black T, D'Arcy S, Weis T, Russell JK (eds) A line in the tar sands: Struggles for environmental justice. PM Press, Oakland, pp 1–20
6. Black T, D'Arcy S, Weis T, Russell JK (eds) (2014) A line in the tar sands: Struggles for environmental justice. PM Press, Oakland
7. Brave Bull Allard L (2016). Why the founder of standing rock Sioux camp can't forget the Whitestone massacre. Yes Magazine. Retrieved from http://www.yesmagazine.org/people-power/why-the-founder-of-standing-rock-sioux-camp-cant-forget-the-whitestone-massacre-20160903
8. Carney N, Kulick A (2016) Rethinking academia and social justice: Reflections from emerging scholars. Berkeley J Sociology, 60. Retrieved from http://berkeleyjournal.org/2016/04/rethinking-academia-and-social-justice-reflections-from-emerging-scholars/
9. Coulthard G, Simpson LB (2016) Grounded normativity/ Place-based solidarity. American Quarterly 68(2):249–255. https://doi.org/10.1353/aq.2016.0038

10. Crampton L (2012) Race and class privilege in the environmental movement. Pachamama Alliance. Retrieved from https://news.pachamama.org/news/race-and-class-privilege-in-the-environmental-movement
11. Crenshaw K (1989) Demarginalizing the intersection of race and sex: A black feminist critique of antidiscrimination doctrine, feminist theory and antiracist politics. The University of Chicago Legal Forum 140:139–167
12. Dayaneni G (2009) Climate justice in the US. In: Brand U, Bullard N, Lander E, Mueller T (eds) Contours of climate justice: Ideas for shaping new climate and energy politics, vol 6. Dag Hammarskjöld Foundation, Uppsala, Sweden, pp 80–85
13. Elbein S (2017, February 5) The youth group that launched a movement at Standing Rock. New York Times. Retrieved from https://www.nytimes.com/2017/01/31/magazine/the-youth-group-that-launched-a-movement-at-standing-rock.html
14. Enzinna W (2017, January/February) "I didn't come here to lose": How a movement was born at Standing Rock. Mother Jones. Retrieved from http://www.motherjones.com/politics/2016/12/dakota-access-pipeline-standing-rock-oil-water-protest/#
15. Foran J, Gray S, Grosse C, LeQuesne T (2018). This will change everything: Teaching the climate crisis. Transformations 28(2): 126–147 https://www.jstor.org/stable/10.5325/trajincschped.28.2.0126
16. Fox J, Spione J, Dewey M (Writers) (2017). Awake: A dream from Standing Rock [DVD]. United States: Digital Smoke Signals and International WOW Company
17. Grosse C (2017) Megaloads and mobilization: The rural people of Idaho stand against big oil. Case Studies in the Environment 1(1):1–7. https://doi.org/10.1525/cse.2017.sc.450285
18. Grosse C (2017b) Working across lines: Resisting extreme energy extraction in Idaho and California. (Doctoral dissertation), Retrieved from https://escholarship.org/uc/item/33g291g0 (ProQuest ID: Grosse_ucsb_0035D_13439. Merritt ID: ark:/13030/m5k69gzg)
19. Haraway D (1988) Situated knowledges: The science question in feminism and the privilege of partial perspective. Feminist Studies 14(3):575–599
20. Hawken P (ed) (2016) Drawdown: The most comprehensive plan ever proposed to reverse global warming. Penguin Books, New York
21. Honor the Earth (n.d.). Treaty rights and oil pipelines: What you need to know [PDF Fact Sheet]. Retrieved from https://static1.squarespace.com/static/58a3c10abebafb5c4b3293ac/t/5b8ee947aa4a99a54b5d275e/1536092488649/factsheet+TREATY+RIGHTS.pdf
22. hooks b (1984) Feminist theory: From margin to center. South End Press, Boston
23. Irving D (2014) Waking up white: And finding myself in the story of race. Elephant Room Press, Cambridge, MA
24. Kemble R (Producer), Moore P (Director) (2018) The eagle and the condor — From Standing Rock with love [Film]. United States
25. LaDuke W (1999) All our relations: Native struggles for land and life. Haymarket Books, Chicago
26. Leopold A, Schwartz CW (1968) A Sand County Almanac, and sketches here and there. Oxford University Press, London
27. McIntosh P (1989) July/August) White privilege: Unpacking the invisible knapsack. Peace and freedom 1989:10–12
28. Moore H, Russell JK (2011) Organizing cools the planet: Tools and reflections to navigate the climate crisis. PM Press, Oakland, CA
29. Painter NI (2010) The history of white people. W.W. Norton & Company, New York
30. Patterson DASW (Adelvunegv Waya), Bulter-Barnes ST (2015) Impact of the academic-social context of American Indian/Alaska Native students' academic performance. Washington University J American Indian & Alaska Native Health, 1(1): 1–30
31. Pickett K (Writer) (2017) First daughter and the black snake [Film]. United States: Pickett Pictures
32. Simpson LB (2014) Land as pedagogy: Nishnaabeg intelligence and rebellious transformation. Decolonization: Indigeneity, education & society 3(3): 1–25

33. Simpson, LB (2017) As we have always done: Indigenous freedom through radical resistance University of Minnesota Press, Minneapolis
34. Smith LT (2012) Decolonizing methodologies: Research and Indigenous peoples, 2nd edn. Zed, London
35. Solnit R (2016) Hope in the dark: Untold histories, wild possibilities, 3rd edn. Haymarket Books, Chicago
36. Thomas-Muller C (2014) The rise of the Native rights–based strategic framework. In: D'Arcy S, Weis T, Russell JK (eds) Black T. Struggles for environmental justice, PM Press Oakland, A line in the tar sands, pp 240–252
37. Whyte KP (2017) The Dakota Access Pipeline, environmental injustice, and US colonialism. Red Ink: An International J Indigenous Literature, Arts, & Humanities 19(1): 154–169
38. Wildcat DR (2009) Red alert!. Saving the planet with Indigenous knowledge, Fulcrum, Golden
39. Windchief S, Joseph DH (2015) The act of claiming higher education as Indigenous space: American Indian/Alaska Native examples. Diaspora, Indigenous, and Minority Education 9(4):267–283
40. Woolford A (2015) This benevolent experiment: Indigenous boarding schools, genocide, and redress in Canada and the United States. University of Nebraska Press, Lincoln

Index

A
Academic colonization, 36
Activism, 36, 37, 42, 44, 46, 50, 56, 68, 84, 119, 143, 146, 147, 150–153
Addison Natural Gas Project (ANGP), 6, 34, 35, 37–39, 45, 49, 52
Administrator, 3, 51, 58, 66, 144, 145
Advocate, 59, 61, 63, 114
Appalachia, 131
Appalachian Trail (AT), 15, 19, 108, 109, 113, 116, 117, 124–127
Appalachian Trail Conservancy (ATC), 19, 21, 29, 120, 122, 126
Assessment, 6, 14–16, 23, 25, 26, 28, 30, 31, 37, 60, 76, 94
Atlantic Coast Pipeline (ACP), 2, 24, 110, 115–117, 119, 120, 125, 129, 131
Autobiographical/autobiography, 35

B
Bakken shale oil, 56
Biodiversity, 78
Biodiversity protection, 36
Biofuelwatch, 36, 45, 47
Blue Ridge Land Conservancy, 21
Blue Ridge Parkway (BRP), 15, 113

C
Campus, 3, 5, 6, 16, 25, 49–51, 58, 60, 65, 68, 78, 92, 144–146, 152
Case study, 6, 14, 16, 18, 21, 23, 27–30, 57, 60–64, 67, 68, 92, 93, 96, 101, 108
Citizenship, 18, 20, 25, 36

Civil rights movement, 47
Climate change, 14, 21, 25, 26, 39, 78, 94, 99, 131, 145–151
Climate justice, 36, 143, 147, 148, 153
Coal, 97
Coalition movements, 88
Co-curricular, 14, 18, 19, 21–23, 25–27, 29–31
College of Saint Benedict and Saint John's University (CSBSJU), 144–146
Colonialism, 153
Colorado, 6, 92, 96, 97
Commitment, 3, 51, 83, 110, 153
Communication
  critical, 92, 93
  environmental, 92–94
  nonverbal, 100
Community
  -based learning, 6, 57, 62
  campus, 2, 5
  -engaged pedagogy, 1, 14
  engagement, 2–4, 20, 27, 57, 62, 150
  organizer, 4, 5, 7, 68
  organizing, 36
Corporate university, 45
Corporation, 1, 48, 110
Critical communication pedagogy (CCP), 93, 95
Critical pedagogy, 2, 95
Critical thinking, 2, 45, 148
Cross-disciplinary, 7
Cultural practices, 14
Culture, 17, 36, 37, 151, 153

D

Dakota Access Pipeline (DAP), 66, 99
Decolonization, 43
Dialogue, 6, 49, 93, 95, 101
Disciplinary, 4, 19, 23, 27, 28, 30, 101
Documentaries, 63

E

Earth Sciences, 6
Eminent domain, 7, 15, 76, 77, 114, 119, 120, 127
Empathy, 6, 14, 27, 29, 148–150
Endangered species, 40, 61, 108, 113, 127, 128
Energy development, 2, 3, 5, 67, 97
Energy transition, 5
Engaged learning, 75
Engagement
 civic, 18, 22, 26, 56–58, 66–68, 83
 stakeholder, 3, 18, 21
Environmental assessment, 77
Environmental communication, 92–94
Environmental Defense Fund, 36
Environmental education, 58
Environmental ethics, 21
Environmental Humanities, 6
Environmental Impact Statement (EIS), 16, 20
Environmental justice, 2, 3, 15, 28, 31, 43, 56–59, 96, 97, 102, 147
Environmental politics, 78, 101
Environmental Politics course, 75, 78
Environmental studies program (ENVS), 14, 16, 17, 21
EQT Corporation, 108
Ethics, 4, 18, 19, 23
Experience, 3, 6, 19, 21, 27, 29–31, 35, 37, 38, 42, 47, 49, 50, 55, 56, 58, 59, 62–64, 66–68, 76, 79, 84, 85, 92, 94, 95, 101, 111, 143–146, 148, 150
Experiential learning, 27, 78, 143
Explosion, 15, 21, 61, 64, 77, 121
 pipeline, 15
Export
 energy, 108, 114, 131
Extraction, 5, 18, 39, 97, 112, 148
Extractive industry, 96
Extractive messaging, 92

F

Facilitated discussion, 23, 34, 150

Faculty, 3, 4, 6, 7, 16, 18, 23, 27, 29, 36, 37, 51, 58, 59, 62, 68, 109, 144–146
Federal Energy Regulatory Commission (FERC), 15, 20, 21, 76, 77, 79, 81, 84, 109, 111–116, 118, 120–122, 124–129, 132
Federal government, 15, 77, 84, 144
Fieldwork methods, 75
Forum
 community, 20
 public, 16, 21, 92
Fossil fuel, 3, 7, 34, 35, 44, 46, 52, 114, 119, 143, 148
Fracked, 2, 6, 15, 34, 36, 44, 49, 142
Fracking, 26, 39, 44, 96, 99, 101, 108, 110, 114, 117, 120, 125, 131

G

Geographic Information Systems (GIS), 6, 21, 58–61, 63, 65
 course, 21
Geography, 37, 58, 60, 122
Geology, 18, 121
Golden-winged warbler, 35, 38–40
Greenwashing, 46, 49, 51

H

Hazards, pipeline, 121
Hearing, 38, 46, 82, 92, 93, 96–101, 109, 115, 118, 122–125, 128, 146
 court, 109, 122
 public, 6, 7, 92, 93, 96–98, 100, 101, 109, 115, 124, 125
Hidden curriculum, viii
Higher Education Institution (HEI) Politics, 35, 44
Historical forces, 21
History, 4, 23, 49, 58, 67, 81, 92, 96, 112, 121, 141, 146, 151, 152
 environmental, 50, 108
Hydrology, 15, 130
Hydro Quebec, 46

I

Identity, 29, 85, 93, 101, 142
Ideology, 19, 93
 environmental, 19
Indigenous education, ix
Indigenous pedagogy, 141
Infrastructure/infrastructuring, 2, 7, 34, 36, 46, 52, 61, 78, 92, 114, 148, 150

Index

Institutions, 14, 16, 18, 37, 45, 46, 60, 92, 142–144, 146, 148, 152
Interdisciplinary, 3, 4, 14, 16–18, 23, 30, 31, 68
Interns, 36
Intervention, 2, 34
Interview, 19, 29, 30, 78, 79, 81, 126, 149
Invasion, 145

J
Jobs, 3, 15, 39, 48, 82, 112, 114
    pipeline, 82

K
Keystone XL pipeline, 118, 131, 148, 149

L
Land education, 144
Law, 63, 67, 96, 100, 112, 113, 117, 127, 128
Learning, 2, 3, 6, 14, 17–19, 21, 27, 31, 34–36, 44, 48, 55, 57, 61, 62, 67, 68, 76, 78, 79, 85, 142, 143, 152
    community-based, 2, 6, 57, 62
Learning outcomes
    course, 20
    program, 23
    student, 76
Legal, 21, 28, 47, 61, 66, 67, 77, 94, 98, 99, 109, 110, 113, 119, 123, 124, 129, 131
Line 3 pipeline, 145, 149, 151, 152
Lobby, 18, 39, 58, 84
    coal, 97
    fossil fuel, 7
    natural gas, 21
Local environments, 1, 20, 141
Local government, 7, 63, 99, 109, 113, 118, 122
Local knowledge, 7, 56, 67, 94, 95
Local resistance, 1

M
Map, 16, 21, 37, 44, 56, 61, 63–65, 80, 122, 123
Media, 17, 48, 60, 62, 63, 109, 110, 115, 119, 124–127, 130, 146, 148
Minnesota, 7, 143–145, 148, 149, 151
Missions, 16
Mobilization, 2, 147

Mountain Valley Pipeline (MVP), 1, 2, 6, 7, 14–31, 107–132
Movement, 19, 35, 41, 42, 44, 46, 47, 75, 76, 79, 80, 85, 96, 142, 144–147, 153
    social, 6, 37, 75, 80, 143, 150
Multi-Logue, 34, 35, 38, 49, 52

N
Narratives, 40, 96, 101
Native nations, 7, 143–146, 148, 152
Natural gas, 1, 2, 6, 15, 21, 36, 75, 76, 78, 79, 108, 110–114, 117, 118, 121, 124, 129, 131, 132
Natural resource, 99
Nested case study, 14
New Jersey, 6, 26, 56–59, 61, 63–66, 76, 77, 81

O
Opposition movement, 34, 75, 80, 85
Oppression, 41, 43, 141–143
350.org, 35, 51, 64
Organizer, community, 4, 5, 7, 36, 68

P
Pedagogy
    case study, 18, 27, 28
    community-engaged, 1, 14
    pipeline, 2–4, 6, 7, 35, 92, 101, 142, 143, 146–148, 150–152
Pedagogy of the oppressed, 2
PennEast Pipeline, 6, 76–78, 81, 84–86
Pilgrims Pipelines, 6, 55, 57, 59–61, 63–67
Pipeline
    crude oil, 56
    energy, 1, 2
    fracked gas, 36
    gas, 36, 128
    liquefied natural gas, 6, 131
    liquid natural gas, 15
    natural gas, 75
    oil, 26, 57, 142
    tar sands oil, 7, 143, 148, 151
Place-based learning, 2, 27, 68
Police brutality, 49
Policy
    energy, 95, 126
    environmental, 6, 59, 94
    -makers, 57, 95
Political ecology, 6
    feminist, 6

Politics, 3, 7, 18, 37, 57, 66, 83, 84, 92, 114, 144
  environmental, 75, 78, 101
Positionality, 3, 142, 143, 147–149, 153
Power relations, 4, 102
Pro-environmental behavior, 40
Property, private, 7, 77, 110, 114, 119, 125, 127, 128, 131
Property rights, 127, 128, 130
Protect Geprags Park, 35, 36, 39, 40, 47, 48
Protest, 3, 6, 7, 19, 38, 66, 79, 97, 100, 108, 109, 121, 125
  pipeline, 3, 16, 51
Public health, 6, 15, 55, 58–60, 62, 63, 67, 68, 94, 97, 98, 101, 102, 120, 122, 123
Public pedagogy, 93

Q
Qualitative research, 80

R
Regulations, 61, 66, 77, 97, 98, 115, 123
Resistance, 7, 16, 34–36, 43, 49, 52, 56, 66, 77, 100, 143, 148, 150
  pipeline, 38, 52, 151, 153
Resource curse, 2, 18
Rider University, 75, 83, 85
Roanoke Appalachian Trail Club (RATC), 21, 120, 124
Roanoke College, 5, 6, 16, 17, 22, 28, 31, 119, 120, 124, 125

S
Sacrifice zone, 2, 18
Safety, 2, 21, 76, 85, 115
  pipeline, 34, 76
Scale, 22, 25, 55, 81
Scientific knowledge, 84, 92, 94
Service learning, 41, 58, 60
Service learning course, 41, 50, 52
Settler colonialism, 34, 141, 142, 144
Shareholders meeting, 38
Social justice, 2, 143, 147, 152
Social justice education, 142
Socioeconomic privilege, 28
Sovereignty, 7, 143, 148, 151, 152
  native nations, 7, 143

Spotted owl, 40
Stakeholder, 1, 3, 15, 16, 18–21, 23, 27–31, 80, 95, 149
Stakeholder politics, 18
Standing Rock, 1, 42, 44, 147
Standing Rock protests, 97, 99
State government, 130
Story, 28, 43, 49, 56, 58, 65, 68, 109, 125, 126, 142
Student, 2–7, 14, 16–31, 36–38, 40–47, 49–52, 56, 58, 60–62, 64, 65, 67, 68, 75, 76, 78–85, 92, 93, 96–102, 108, 109, 124, 125, 130, 142–150, 152, 153
Student response, 16
Subjectivity, 19, 94
Survival, 64, 127, 142, 153

T
Tar sands, 56, 148
Teamwork, 62
Technical expertise, 93, 95, 96, 98–100
Technical knowledge, 96, 99–101
Technofixes, 45
TransCanada Mainline, 34
Transdisciplinary, 2
Tribal school, 143, 152

U
University of Vermont (UVM), 35–37, 41–43, 46, 49–51
Utility company, -ies, 123

V
Vermont, 6, 34–42, 44, 45, 47, 49, 50, 52
Vermont natural gas, 6, 36
Virginia, 6, 7, 15, 21, 26, 107, 108, 110–114, 116–118, 120, 121, 124–130
Visualization, 21

W
Water quality, 16, 56, 116, 121, 123, 127, 130
West Virginia, 15, 108, 113, 117, 129, 130
Wisconsin, 40, 41
Workers, 40, 82, 118
  imported, 40
  pipeline, 112